OPTICAL FIBERS FOR DESIGNING MULTIPLE APPLICATIONS

OPTICAL FIBERS FOR DESIGNING MULTIPLE APPLICATIONS

DEEPA VENKITESH
NIMISH DIXIT
AND
R. VIJAYA

Nova Science Publishers, Inc.
New York

Copyright © 2009 by Nova Science Publishers, Inc.

All rights reserved. No part of this book may be reproduced, stored in a retrieval system or transmitted in any form or by any means: electronic, electrostatic, magnetic, tape, mechanical photocopying, recording or otherwise without the written permission of the Publisher.

For permission to use material from this book please contact us:
Telephone 631-231-7269; Fax 631-231-8175
Web Site: http://www.novapublishers.com

NOTICE TO THE READER

The Publisher has taken reasonable care in the preparation of this book, but makes no expressed or implied warranty of any kind and assumes no responsibility for any errors or omissions. No liability is assumed for incidental or consequential damages in connection with or arising out of information contained in this book. The Publisher shall not be liable for any special, consequential, or exemplary damages resulting, in whole or in part, from the readers' use of, or reliance upon, this material.

Independent verification should be sought for any data, advice or recommendations contained in this book. In addition, no responsibility is assumed by the publisher for any injury and/or damage to persons or property arising from any methods, products, instructions, ideas or otherwise contained in this publication.

This publication is designed to provide accurate and authoritative information with regard to the subject matter covered herein. It is sold with the clear understanding that the Publisher is not engaged in rendering legal or any other professional services. If legal or any other expert assistance is required, the services of a competent person should be sought. FROM A DECLARATION OF PARTICIPANTS JOINTLY ADOPTED BY A COMMITTEE OF THE AMERICAN BAR ASSOCIATION AND A COMMITTEE OF PUBLISHERS.

LIBRARY OF CONGRESS CATALOGING-IN-PUBLICATION DATA

ISBN: 978-1-60692-782-3

Available upon request

Published by Nova Science Publishers, Inc. New York

CONTENTS

Preface		vii
Chapter 1	Introduction	1
Chapter 2	Origin of Third Order Nonlinearity	5
Chapter 3	Nonlinear Refractive Index and the Consequent Effects in Optical Fibers	7
Chapter 4	Nonlinear Pulse Propagation	13
Chapter 5	Split-Step Fourier Transform Technique (SSFT)	17
Chapter 6	Propagation Characteristics of Pulses in Different Regimes of Operation	23
Chapter 7	Specific Applications of Third Order Nonlinearity	33
Chapter 8	Summary	67
Acknowledgements		69
References		71
Index		79

PREFACE

Nonlinear optical phenomena are easily observable in high-power fiber-optic communication systems. The generation of new frequencies due to these phenomena may be used to design different applications. Four wave mixing, which is conventionally viewed as a deterrent to multi-channel wavelength-division-multiplexed propagation of signals in communication systems, can be used favorably to design optical sources with special characteristics such as multi-wavelength, frequency comb output and supercontinuum. These have applications in widely different fields like communication and medicine.

Chapter 1

INTRODUCTION

Nonlinear optics is the study of phenomena that occur as a consequence of the modification of the optical properties of a material system by the presence of light. These modifications in the response of the material depend on the strength of the optical field in a nonlinear manner. Optical fibers are ideal for nonlinear interactions because they provide a very strong confinement of field over long propagation distances. In the case of bulk nonlinear optics, the high intensity required for the onset of nonlinear interaction is often obtained by focusing the laser beams. The interaction length is however limited as the focusing becomes tighter, due to diffraction limits. In the case of optical fibers, since the beam is confined to a dimension of the order of wavelength of light, the nonlinear interaction length is limited only by the linear attenuation in the fiber. The interaction lengths can be of the order of kilometers, thus making it an excellent medium to observe nonlinear effects. There are numerous discussions on the origin of nonlinear effects and their consequences available in the literature [1-6].

The steady state nonlinear response of a medium is governed by nonlinear polarizations and nonlinear susceptibilities. These are the characteristic properties of a medium and they depend on the detailed electronic and molecular structure of the medium. The lowest order nonlinear effects in fibers are due to the third order nonlinear susceptibility, $\chi^{(3)}$, under the electric dipole approximation. Both the real and imaginary components of $\chi^{(3)}$ lead to interesting phenomena. The real part gives rise to third harmonic generation and intensity dependent refractive index with its consequent effects like self phase modulation, cross phase modulation and four wave mixing, while the imaginary part gives rise to nonlinear scattering effects like the stimulated Raman and Brillouin scattering.

Third harmonic generation is the process in which a frequency, three times that of the incident light is generated through the nonlinear interaction. Its efficiency is quite low in optical fibers. Self phase modulation (SPM) is the phenomenon by which the phase of a transmitted pulse is altered by its own intensity. This, together with the dispersion in the fiber can change the temporal width of the pulse. Cross phase modulation (XPM) is similar to SPM, but it occurs when two pulses propagating simultaneously in a fiber overlap, and the refractive index modified by the one affects the other and vice versa. Four wave mixing (FWM) is the process by which the fields at different frequencies propagating simultaneously through the medium interact due to the third order nonlinearity, leading to the generation of new frequencies. Stimulated Raman and Brillouin scattering are the consequences of nonlinear scattering phenomena, which also result in the generation of new frequencies.

All the nonlinear effects discussed above are, in general, detrimental to communication systems which employ optical fibers. In the case of a linear communication system, the capacity of a transmitting system (the maximum possible bit rate allowed in the system) is the product of the usable bandwidth of the channel and its spectral efficiency. The spectral efficiency of a linear system depends only on the signal-to-noise (S/N) ratio, and according to Shannon's formula [6], the capacity can be increased by increasing the S/N ratio. For optical communication in the S, C and L bands, the spectral bandwidth is about 90 nm and the corresponding usable frequency bandwidth is about 11 THz. Assuming a S/N ratio of 25 dB, the capacity of the system is about 91 terabits/s. When the signal power is increased, the efficiency of the nonlinear processes discussed above starts increasing, and hence adds to the noise in the system. Thus, for a nonlinear system, there exists an optimal signal power beyond which the noise limits the channel capacity.

The communication system is affected by the various nonlinear effects in the following way. SPM alters the phase of the propagating pulse, and hence modifies its spectral width. Shorter pulses, which are richer in spectral content are affected more. The extent of alterations in the pulse quality due to SPM depends strongly on the magnitude and nature of group velocity dispersion in the fiber. Along with the group velocity dispersion, SPM decides the permissible bit rate of transmission in a system. The effects of this process can be compensated through an intelligent system design. The intensity of one channel affects the phase of the pulse in a simultaneously propagating second channel due to XPM. The extent of XPM generated is a function of the walk-off between the two pulses. XPM leads to broadening of pulses and adds to additional noise in the system, thus reducing the S/N ratio. The frequency broadening in XPM is found to be twice as strong as that due to SPM.

FWM is a major obstacle in wavelength division multiplexed systems, where the adjacent channels of the system mix to generate new frequencies and the generated frequencies coherently superpose on the receiver, which increases the S/N ratio. The fact that FWM can generate altogether new frequencies makes it distinctly different from XPM. For FWM to occur, the phase matching conditions need to be satisfied. The effective phase difference due to dispersive and nonlinear effects need to be considered while deriving the phase matching conditions. Mixing products coinciding with the original WDM channels, result in the degradation of system performance.

Stimulated Brillouin scattering (SBS) is highly detrimental to communication systems since it limits the maximum power that can be effectively transported. Since the scattered wave propagates in the backward direction, it can cause destabilization in the source. Once the launched power exceeds the Brillouin threshold, it leads to fluctuations in the transmitted power. Since SBS threshold depends on the line width of the pump, it is usually countered in commercial systems by enhancing the line width of the pump by phase modulation. Practically, the achievable spectral efficiency and bit rates also depend on the modulation format and the detection technique.

Nonlinear effects in fibers have also proved to be very useful in many applications. They offer a variety of possibilities in ultra fast optical switching, optical amplification and optical regeneration. The objective of this chapter is to present the utility of optical nonlinearities in fibers for designing special applications. The correct choice of SPM and dispersion can lead to the generation of optical solitons. Transmission of optical signals over trans-oceanic distances without distortion is possible with the use of solitonic pulses. Four wave mixing is found to be extremely useful for parametric amplification and wavelength conversion. Raman amplification has emerged as one of the favourite schemes for distributed amplification, using which, terabit transmission rates are demonstrated. Enhanced bandwidth and the absence of a special fiber make Raman amplifiers superior to conventional erbium doped fiber amplifiers. Nonlinear effects in fiber can be enhanced with the design of speciality fibers like the highly nonlinear fibers and nonlinear photonic crystal fibers. Brillouin effect is utilized to design multi-wavelength optical sources and for Brillouin amplification. Brillouin amplifiers require a very low pump power and are found to be useful for specific applications which require a narrow gain bandwidth. Raman and Brillouin effects are also widely used for optical fiber sensor applications.

This book discusses in detail, the various aspects of third-order nonlinearity in optical fibers for designing multiple applications. Section 2 presents the origin of third order susceptibility at a molecular level and the consequences. Though the absolute value of nonlinear index coefficient is fairly low for silica fibers, the small area of interaction leads to a large value of nonlinear index parameter, and the consequent effects are further analyzed in section 3. The intricacies involved in the numerical solution of the nonlinear Schrödinger equation, which governs the pulse propagation through the nonlinear medium, using the split step Fourier Transform are discussed at length in sections 4 and 5. The results of the solution in different regimes of operation are discussed qualitatively for Gaussian pulse propagation, followed by a detailed numerical analysis for the propagation of a beat signal through a fiber in section 6. Some specific applications of nonlinear effects are discussed subsequently in section 7. One of the prominent applications is in the design of a multiwavelength source for communication systems, where an amplified beat signal is propagated through a low dispersion fiber, resulting in enhanced nonlinear effects. A detailed theory of degenerate four wave mixing with analytical and experimental results is discussed next with some of its applications. The origin of Stimulated Brillouin scattering is examined and the numerical solution of the coupled differential equation for the propagation of pump and Stokes wave is presented. The experimental technique to measure the Brillouin threshold is explained and some experimental results are demonstrated. Raman scattering and the applications of Stimulated Raman scattering are presented followed by a very brief discussion on soliton lasers and super continuum generation using photonic crystal fibers.

Chapter 2

ORIGIN OF THIRD ORDER NONLINEARITY

When a high frequency electric field (E(t)) of an electromagnetic wave is incident on a dielectric, the resulting distortion in the electron cloud surrounding the atoms in the dielectric results in a field-induced dipole moment, which in turn is a source of electromagnetic radiation. The secondary wave thus radiated by the atom superimposes on the primary one and this results in the propagation of an optical field in the medium at a microscopic level. The algebraic sum of these distinct microscopic field induced dipole moments per unit volume, referred to as polarization P(t), is a macroscopic measure of the field-induced changes. In the case of conventional linear optics, the electron clouds move in a parabolic potential, and hence the induced polarization is linearly proportional to the strength of the electric field. However, as the field strength increases, the restoring force on the electron cloud due to the Coulomb attraction is no longer linear and this effect can be included by approximating the restoring force by a power series. Consequently, the polarization can also be expanded in a power series as

$$P = \varepsilon_0 \left(\chi^{(1)} E + \chi^{(2)} EE + \chi^{(3)} EEE + ... \right) = P^{(1)} + P^{(2)} + P^{(3)} + ... \tag{1}$$

where $\chi^{(1)}$, $\chi^{(2)}$, $\chi^{(3)}$ are the linear, second and third order susceptibilities respectively.

Considering that the electric field is a vector quantity, $\chi^{(1)}$ is a second rank tensor, $\chi^{(2)}$ is a third rank tensor and so on. $P^{(1)}$ determines the linear part of polarization, while $P^{(2)}$ and $P^{(3)}$ represent the nonlinear polarization of second and third order respectively [3]. While writing eqn (1) in that form, it is assumed that the polarization at time t depends only on the instantaneous value of the electric field strength. It also assumes that the nonlinear optical response is instantaneous, which is true for bound-electron nonlinearities. Another assumption is the locality, which implies that the nonlinear polarization at a given point in space depends on the magnitude of electric field at that point. This condition is not always satisfied, for example, in electrostrictive nonlinearities. However, eqn (1) can be used to describe the various third order effects that are local and non resonant. In general, the nonlinear susceptibilities also depend on the frequencies of the applied fields.

It can be shown that, for centro-symmetric materials (which possess a center of inversion), the $\chi^{(2)}$ nonlinear susceptibility vanishes to zero. Optical fibers used in telecommunication are made of silica glass in an amorphous state. Hence, the second order susceptibility and its effects, namely second harmonic generation and sum frequency generation, are absent [1,3]. There are exceptions to this, due to the contributions from the electric quadrupole and magnetic dipole moments in the material and also due to defects or color centers inside the fiber core.

The nonlinear optical effects in the telecommunications range (1.3 μm – 1.6 μm) are predominantly due to $\chi^{(3)}$. Out of the 81 elements of $\chi^{(3)}$, only 21 are non zero due to symmetry considerations. In the case of optical fibers, these residual elements can be written in terms of three elements. Neglecting the frequency dependence of the third order susceptibility as well as the Raman contribution to susceptibility, all these components can be written in terms of a single term, which can be considered as a constant [3,7]. The term involving $\chi^{(3)}$ has three optical fields interacting to produce a fourth field, and hence $\chi^{(3)}$ interaction is essentially a four-photon process. It is primarily responsible for intensity dependent refractive index (optical Kerr effect), the self phase modulation (SPM), cross-phase modulation (XPM), four wave mixing (FWM), the Stimulated Raman Scattering (SRS) and the Stimulated Brillouin Scattering (SBS). The occurrence of all these phenomena is usually considered spurious and contributing to the noise in an optical communication system. However, each of these effects can be optimally utilized towards specific needs. One such application is the generation of new frequencies in optical fibers. Each of these effects is discussed in brief in the following sections.

Chapter 3

NONLINEAR REFRACTIVE INDEX AND THE CONSEQUENT EFFECTS IN OPTICAL FIBERS

One of the most important nonlinear effects based on third order nonlinear susceptibility is the nonlinear refractive index. The origin of nonlinear refractive index can be understood by the following discussion. Consider a plane monochromatic wave propagating in a dielectric medium, given by

$$E(z,t) = \frac{1}{2}\left(\hat{E}(z)e^{-j\omega t} + c.c\right) \tag{2}$$

The linear wave equation governing the propagation of electromagnetic waves in a dielectric can be derived from the Maxwell's equations as

$$\Delta E = \frac{1}{c^2}\frac{\partial^2 E}{\partial t^2} + \frac{1}{\varepsilon_0 c^2}\frac{\partial^2 P}{\partial t^2} \tag{3}$$

where Δ is the Laplacian operator. Considering the propagation of the waves in the same direction, the vectorial nature of the fields is not considered while writing eqn (3). With linear polarization, $P_L = \varepsilon_0 \chi^{(1)} E$, eqn (3) becomes

$$\Delta E = \frac{1}{c^2}\frac{\partial^2 E}{\partial t^2}\left(1 + \chi^{(1)}\right) \tag{4}$$

The linear refractive index, n, is defined using equation,

$$n = \sqrt{1+\chi^{(1)}} = \sqrt{\varepsilon_r} \qquad (5)$$

where ε_r is the material dependent relative permittivity. In general, the relative permittivity is complex and is a tensor of second rank, and consequently, the linear refractive index is also complex. In terms of the linear refractive index, eqn (4) can be rewritten as,

$$\Delta E = \frac{n^2}{c^2}\frac{\partial^2 E}{\partial t^2} \qquad (6)$$

When the intensity of the incident wave is high enough to invoke nonlinear responses, then the polarization of an inversion symmetric material (like silica) would be given by

$$P = \varepsilon_0 \left(\chi^{(1)} E + \chi^{(3)} EEE + ... \right) \qquad (7)$$

The lowest order nonlinear polarization is given by

$$P^{(3)} = \chi^{(3)} EEE \qquad (8)$$

For a plane wave, it can be re-derived by substituting eqn (2) in eqn (8) as

$$P^{(3)} = \frac{1}{8}\varepsilon_0 \chi^{(3)} \left(\left(\hat{E}^3 e^{-j3\omega t} + c.c \right) + 3|\hat{E}|^2 \left(\hat{E} e^{-j\omega t} + c.c \right) \right) \qquad (9)$$

The first term in the above equation is due to an oscillating field at the third harmonic frequency, 3ω and this term can contribute significantly to the third order nonlinear polarization only when phase matching conditions are satisfied. For optical fibers, this term is usually negligible, which results in

$$P^{(3)} = \frac{3}{4}\varepsilon_0 \chi^{(3)} |\hat{E}|^2 E \qquad (10)$$

Substituting eqn (10) in the wave equation (eqn (3)),

$$\Delta E = \frac{1}{c^2}\frac{\partial^2 E}{\partial t^2} + \frac{1}{\varepsilon_0 c^2}\frac{\partial^2}{\partial t^2}\left[\varepsilon_0 \chi^{(1)} E + \frac{3}{4}\varepsilon_0 \chi^{(3)}|\hat{E}|^2 E\right]$$

$$= \frac{1}{c^2}\frac{\partial^2 E}{\partial t^2}\left[1 + \chi^{(1)} + \frac{3}{4}\chi^{(3)}|\hat{E}|^2\right] \qquad (11)$$

Comparing eqn (11) with eqn (6), the refractive index of the material has an additional term due to the third order nonlinear susceptibility, and the total refractive index can be written as,

$$\tilde{n} = \sqrt{1 + \chi^{(1)} + \frac{3}{4}\chi^{(3)}|\hat{E}|^2} \qquad (12)$$

Assuming that the nonlinear part of the refractive index is relatively small compared to the linear component, eqn (12) can be expanded as a Taylor series about $\frac{3}{4}\chi^{(3)}|\hat{E}|^2$ resulting in

$$\tilde{n} = n + \frac{3}{8n}\chi^{(3)}|\hat{E}|^2 \qquad (13)$$

with the linear component, n defined according to eqn (5).

The intensity of electromagnetic field represented by eqn (2) is given by,

$$I = \frac{1}{2}\varepsilon_0 cn|\hat{E}|^2 \qquad (14)$$

Using eqn (14), eqn (13) can be rewritten as,

$$\tilde{n} = n + n_2 I \qquad (15)$$

with $n_2 = \dfrac{3\chi^{(3)}}{4\varepsilon_0 c n_0^2} \qquad (16)$

n_2 is generally referred to as the nonlinear index coefficient. Thus, the total refractive index, $\tilde{n}$, consists of a linear part, which is dependent on the frequency, and a nonlinear part, dependent on the intensity. Since the intensity dependence is analogous to the electro-optic Kerr effect, the phenomenon is also called as optical or nonlinear Kerr effect. The nonlinear index coefficient depends on the linear refractive index and the third order susceptibility, which are in turn frequency dependent. The standard value of n_2 for optical fibers in the telecommunication range is $\sim 2.3 \times 10^{-20} \, m^2 W^{-1}$ [8]. This value can be increased with germanium concentration in the core, resulting in highly nonlinear fibers [9]. The value of n_2 is also found to be significantly (about 1000 times) higher in chalcogenide glasses [10]. Though the nonlinear susceptibility and the linear refractive index are used as scalars in the above analysis, they normally depend on the polarization states of the waves and hence the values of n_2 obtained experimentally are assumed to be averaged over all possible polarization states [6].

Even though the numerical value of n_2 is relatively small in communication grade optical silica fibers, the field is concentrated in a small core area in single mode fibers. This makes the various nonlinear effects observable in optical fibers. Since the field strength in the guided mode in a single mode fiber has a specific distribution, the effective area of guidance is calculated by the following overlap integral,

$$A_{eff} = \frac{2\pi \left(\int_0^\infty |E(r)|^2 r \, dr\right)^2}{\int_0^\infty |E(r)|^4 r \, dr} = \frac{2\pi \left(\int_0^\infty I(r) r \, dr\right)^2}{\int_0^\infty I^2(r) r \, dr} \quad (17)$$

where $E(r)$ is the amplitude and $I(r)$ is the intensity of the near-field distribution of the fundamental mode, at a radial distance r from the axis of the fiber. In a single mode step index fiber, the field of the fundamental mode is approximated by a Gaussian function, with a width parameter w, which is a function of the wavelength of propagation (λ). In this case, the effective area is approximated as,

$$A_{eff} = \pi w^2 \quad (18)$$

Typically, the effective area of the fiber is in the range of 20-100 μm^2, depending on the type of fiber. Since a reduced value of the effective area is the key to observing nonlinear effects in fibers, the strength of the nonlinear effect is expressed in terms of the nonlinear parameter, γ, which is defined as

$$\gamma = k_0 \frac{n_2}{A_{eff}} = \frac{2\pi n_2}{\lambda A_{eff}} \tag{19}$$

In a standard single mode fiber (SMF), γ takes values in the range 1-10 W^{-1}km^{-1}. Dispersion shifted fibers have smaller effective areas and hence would have larger nonlinear parameters. The highest nonlinear parameters can be found for specially designed nonlinear fibers like the highly nonlinear fiber (HNLF), where the value of n_2 is also deliberately made larger. Since the field is guided in an optical fiber, the intensity decreases only by the very low attenuation in the fiber and hence, the interaction length can be several kilometers.

The direct consequence of high nonlinear coefficient is the self phase modulation (SPM). SPM occurs whenever a signal having a time varying amplitude is propagated in a nonlinear material. Considering the signal represented by eqn (2), the propagating signal in the medium can be written as

$$E(z,t) = \hat{E}(z) e^{-j[\omega_0 t - [n + n_2 I(z,t)] k_0 z]} \tag{20}$$

$\hat{E}(z)$ is the amplitude of the wave and k_0 is the free space propagation constant. Since the pulse intensity changes with time, the phase of the propagating signal is modulated. The instantaneous phase and its time derivative (frequency) can be written as,

$$\phi(z,t) = -[\omega_0 t - [n + n_2 I(z,t)] k_0 z] \text{ and } \omega' = \omega_0 - n_2 k_0 z \frac{\partial I}{\partial t} \tag{21}$$

The effect of SPM is seen qualitatively from eqns (20) and (21). The instantaneous phase has a nonlinear component, which depends on the intensity of the pulse and the propagated distance. The additional frequency components imposed on the pulse increase the spectral width of the pulse. Also, a frequency sweep or chirp is imposed on the pulse, and the direction of the chirp depends on the direction of $\frac{\partial I}{\partial t}$. Due to the group velocity dispersion, different wavelength components travel with different group velocities resulting in a frequency chirp. The frequency chirp introduced due to SPM can add or subtract from the chirp imposed by linear group velocity dispersion.

Cross phase modulation (XPM) is similar to SPM, except that, two overlapping, but distinguishable pulses, having different frequencies or polarizations are involved. One pulse modulates the index of the medium which leads to the phase modulation of the second pulse. Thus, XPM is a cross-talk mechanism between two channels in those communication systems where intensity modulation is used. No transfer of energy occurs between the channels, and this distinguishes XPM from the other cross talk mechanisms, in which the growth of signal occurs in one channel at the expense of the decay in the feeding channel. Though the XPM process is expected to be twice as strong as SPM, the efficiency of the process is weakened by the fact that pulses of differing frequencies or polarizations are generally group velocity mismatched, and hence cannot maintain their overlap indefinitely.

The effects of SPM and XPM in an optical fiber can be understood quantitatively by solving the Maxwell's equations for pulse propagation through a nonlinear medium. This is discussed in the following section.

Chapter 4

NONLINEAR PULSE PROPAGATION

The study of nonlinear fiber optics involves the propagation of optical pulses inside an optical fiber. When an optical pulse propagates through a fiber, its time and spectral characteristics are affected by three main processes viz., (i) losses associated with the material of the optical fiber, (ii) chromatic dispersion, which arises due to the frequency dependence of the effective index of the mode of the fiber and (iii) optical nonlinear effects, which originate because of the variation of refractive index with the intensity of the pulse.

The basic equation that governs the propagation of an optical pulse in an optical fiber which includes the three effects mentioned above is called the nonlinear Schrödinger equation (NLSE) [5]. This equation can be derived from eqn(11) as

$$\frac{\partial A}{\partial z} = -\frac{\alpha}{2}A - i\frac{\beta_2}{2}\frac{\partial^2 A}{\partial T^2} + \frac{\beta_3}{6}\frac{\partial^3 A}{\partial T^3} + i\gamma\left(|A|^2 A + \frac{i}{\omega_0}\frac{\partial}{\partial T}\left(|A|^2 A\right) - T_R A\frac{\partial |A|^2}{\partial T}\right) \quad (22)$$

In the above equation, $A(t)$ denotes the complex amplitude of the slowly varying optical field. T is the time measured in the frame of reference which is moving with the pulse with group velocity, v_g, so that, $T = t - \left(z/v_g\right)$. The term proportional to α is the loss term and the term proportional to β_2 ($=\frac{d^2\beta}{d\omega^2}$ i.e. second order derivative of the propagation constant β with frequency ω) governs the effect of second order dispersion or group velocity dispersion (GVD), which arises due to the frequency dependence of the effective mode index. The third term represents the contribution due to the third order dispersion, originating from the cubic term in the expansion of propagation constant. This term is important in ultra short pulses and in the case of propagation through a fiber at a wavelength where β_2 is close to zero. The fourth term takes the effect of SPM into account. The term proportional to ω_0^{-1} results from the first derivative of $P^{(3)}$ in the wave equation, and it is responsible for self-steepening and shock formation. The last term has its origin in the delayed Raman response, and it is responsible for the self-frequency shift induced by intra-pulse Raman scattering. The nonlinear response of the medium should include the electronic and vibrational (Raman) contributions. The electronic response is instantaneous while the Raman response is comparatively sluggish. Hence, the nonlinear response function is mathematically written as,

$$R(t) = (1 - f_R)\delta(t) + f_R h_R(t) \tag{23}$$

The electronic response is represented through the Dirac delta function, $\delta(t)$ and f_R represents the fractional contribution of the delayed Raman response to the nonlinear polarization. The Raman response function itself is represented by $h_R(t)$. The sluggish effect of Raman response is incorporated through the term T_R in eqn (22), which is defined as the first moment of the nonlinear response function. This term becomes important and should be included for short pulses with a broad spectrum, where the lower frequency components of the pulse get amplified at the expense of the higher frequency components of the same pulse due to the energy transfer from higher frequency components to the lower frequency components. This phenomenon is called intrapulse Raman scattering. Due to this reason, the pulse spectrum shifts towards the low frequency side as the pulse propagates down the fiber. T_R can be related to the slope of the Raman gain spectrum [11] and its numerical value is deduced experimentally [12] as ~ 3 fs at wavelengths ~ 1.55 μm. For pulses shorter than 1 ps, the Raman gain does not vary linearly over the entire pulse bandwidth and hence, T_R needs to be replaced by an appropriate fitting parameter.

It is important to note that, while arriving at eqn (22), the following assumptions are made:

(1) Only the fundamental mode is assumed to propagate through the fiber. Hence the fiber to be analyzed has to be single moded for the wavelength considered.
(2) The material is perfectly transparent and the wavelength of the applied field is far from any material resonances; i.e, the absorption and nonlinear resonances are neglected.
(3) All scattering effects in the waveguide are neglected. Hence Rayleigh scattering, Rayleigh wing scattering, Raman and Brillouin scattering play no significant role.
(4) The amplitude of the wave packet considered changes only very slowly with respect to its carrier. This is also called the slowly varying envelope approximation.
(5) The fields are linearly polarized and the polarization directions remain the same during propagation. i.e, the birefringence and the vectorial nature of the waves are neglected.
(6) The nonlinearity has no influence on the field components perpendicular to the direction of propagation.
(7) The nonlinearities are to be considered only as perturbations to the linear behavior. i.e, the applied field is much weaker than the inter-atomic field.

In this context, it is useful to define two important length scales, namely, dispersion length (L_D) and nonlinear length (L_{NL}) as [5]

$$L_D = \frac{T_w^2}{|\beta_2|} \qquad L_{NL} = \frac{1}{\gamma P_0} \qquad (24)$$

where T_w is the width of the pulse. It is related to the full-width at half maximum (FWHM) of the pulse by $T_{FWHM} = 1.665 \times T_w$ for a Gaussian pulse and $T_{FWHM} = 1.763 \times T_w$ for the secant hyperbolic pulse. P_0 is the input peak power of the pulse. Dispersion length (L_D) and nonlinear length (L_{NL}) are the lengths over which the dispersive and nonlinear effects become important for the pulse evolution. For a standard single mode fiber, with $\gamma = 2.6 W^{-1} km^{-1}$, and with a GVD parameter $-27.9 ps^2 / km$, the values of dispersion and nonlinear length for different powers and pulse widths are shown in Figure 1.

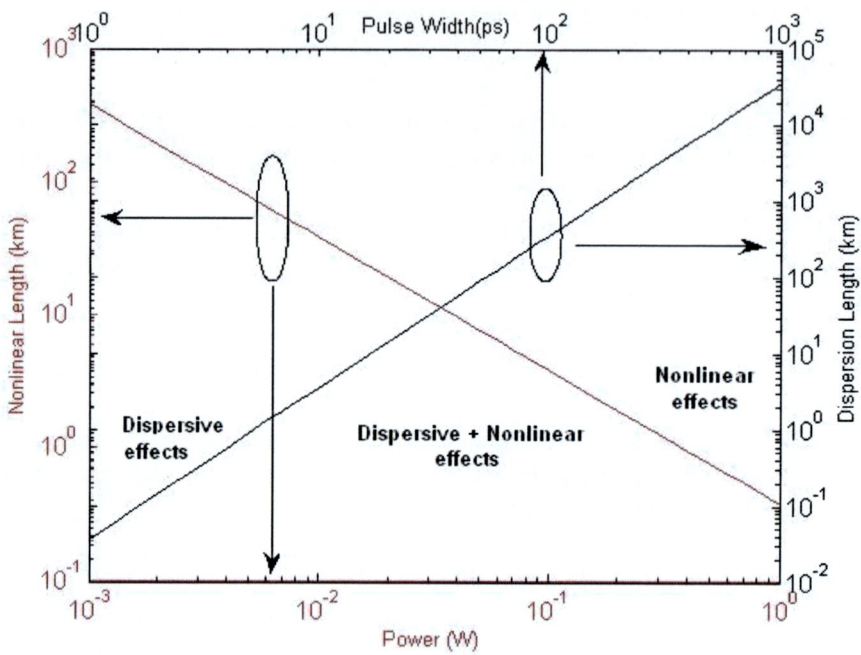

Figure 1. Variation of dispersion length with pulse width and nonlinear length with power, to identify the regimes of operation for pulse propagation in an optical fiber.

It can be interpreted from Figure 1 that, for a length (L) of the fiber, if L<<L_{NL} and L<<L_D, neither dispersive nor non-linear effects play a significant role during pulse propagation. The fiber plays a passive role in this regime and acts as a mere transporter of optical pulses. When L<<L_{NL} and L ≥ L_D, the pulse evolution is governed by GVD and non-linear effects play a relatively minor role. If L<<L_D but L ≥ L_{NL}, the pulse evolution is governed by self phase modulation (SPM) and leads to the spectral broadening of the pulse. Lastly, if L >> L_D and L >> L_{NL}, the dispersion and non-linearity act together as the pulse propagates along the fiber. This condition is significant because, the interplay between GVD and SPM leads to the soliton generation in the anomalous dispersion regime (β_2 <0). In the normal dispersion regime (β_2>0), the GVD and SPM can be used for pulse compression. The following section discusses the numerical solution of the NLSE using the most common technique of split-step Fourier transform (SSFT) method.

Chapter 5

SPLIT-STEP FOURIER TRANSFORM TECHNIQUE (SSFT)

In the case of pulse propagation where both dispersive as well as nonlinear effects dominate, the NLSE has to be solved numerically for a given pulse shape. One of the most commonly used numerical schemes is the split step Fourier method. The NLS equation (22) can be written in the following form [5]

$$\frac{\partial A}{\partial z} = \left(\hat{D} + \hat{N} \right) A \qquad (25)$$

where $\hat{D}$ is the differential operator that takes care of dispersion and attenuation of the material of the optical fiber and $\hat{N}$ is the nonlinear operator that accounts for the effect of fiber nonlinearities on pulse propagation. These operators are given by

$$\hat{D} = -\frac{i\beta_2}{2}\frac{\partial^2}{\partial T^2} + \frac{\beta_3}{6}\frac{\partial^3}{\partial T^3} - \frac{\alpha}{2} \qquad (26)$$

$$\hat{N} = i\gamma \left[|A|^2 + \frac{i}{\omega_0}\frac{1}{A}\frac{\partial}{\partial T}\left(|A|^2 A\right) - T_R \frac{\partial |A|^2}{\partial T^2} \right] \qquad (27)$$

There is a dynamic interplay between chromatic dispersion and optical nonlinearity as the pulse propagates down the fiber. The solution of NLSE using SSFT is based on the assumption that, during propagation over a sufficiently small distance h, dispersive and nonlinear effects act independently. In other words, there are two steps involved in the propagation from z to $z + h$. In the first step, only nonlinearity acts and dispersion is assumed to be zero, whereas in the second step, dispersion acts and nonlinearity is supposed to be zero. Mathematically, this can be expressed as

$$A(z+h,T) \approx \exp\left(h\hat{D}\right)\exp\left(h\hat{N}\right)A(z,T) \qquad (28)$$

The exponential operator $\exp(h\hat{D})$ can be calculated in Fourier domain by using the prescription

$$\exp\left(h\hat{D}\right)B(z,T) = F_T^{-1}\exp\left[h\hat{D}(i\omega)\right]F_T B(z,T) \qquad (29)$$

where F_T denotes the Fourier transform operation, $\hat{D}(i\omega)$ is obtained from equation (26) by replacing the differential operator $\partial/\partial T$ by $i\omega$, and ω is the frequency in the Fourier domain. Thus $\hat{D}(i\omega)$ is just a number in the Fourier domain and it becomes straightforward to evaluate equation (28) by using FFT algorithm. This scheme is accurate to the second order in step size h.

The accuracy of the SSFT can be improved by adopting a different procedure to propagate the optical pulse over one segment z to $z + h$. In this method, the effect of nonlinearity in included in the middle of the segment rather than at the segment boundary. Mathematically, this can be written as

$$A(z,T) \approx \exp\left(\frac{h}{2}\hat{D}\right)\exp\left(\int_z^{z+h}\hat{N}(z')dz'\right)\exp\left(\frac{h}{2}\hat{D}\right)A(z,T) \qquad (30)$$

This method is known as symmetrized SSFT because of the presence of symmetric exponential operators. This scheme is accurate to the third order in step size h [5]. The integral in the middle part is used to include the z dependence of the nonlinear operator. If the step size is very small, this integral can be approximated by the exponential nonlinear operator given in equation (27). To further improve the accuracy, the integral in equation (30) may be calculated using the trapezoidal rule [5],

$$\int_{z}^{z+h} \hat{N}(z')dz' \approx \frac{h}{2}\left[\hat{N}(z) + \hat{N}(z+h)\right] \qquad (31)$$

The implementation of the above equation is not straightforward because $\hat{N}(z+h)$ is not known at the middle segment located at $z + h/2$. Thus, it becomes necessary to follow an iterative procedure that is initiated by replacing $\hat{N}(z+h)$ by $\hat{N}(z)$. Equation (30) is then evaluated to get $A(z+h, T)$ which in turn is used to calculate the new value of $\hat{N}(z+h)$. Two such iterations are enough in practice.

The method discussed in equation (30) is straightforward to implement. But the efficiency of the split-step method depends on both the time domain resolution and on the distribution of step sizes along the fiber; it requires that the step sizes in z and T are selected efficiently to maintain the accuracy. The following are the techniques used to choose the step size in SSFT [13]:

1. CONSTANT STEP SIZE METHOD

This is the simplest and widely used method, in which the accuracy can be improved by increasing the total number of steps. Convergence is checked by simply reducing the step size and ensuring repeatability.

2. NONLINEAR PHASE ROTATION METHOD

This is a variable step size method that is designed specifically for systems in which nonlinearity plays a major role. For a step size h, the effect of nonlinear operator $\hat{N}$ is to increment the phase of the signal by an amount,

$\phi_{NL}(z,T) = \gamma|A|^2 h$. An upper limit, ϕ_{NL}^{max} is imposed on the nonlinear phase increment, and the bound of the step size would be,

$$h \leq \frac{\phi_{NL}^{max}}{\gamma|A|^2} \tag{32}$$

This criterion is specifically useful for simulating soliton propagation and it is widely used in optical fiber transmission simulators.

3. LOGARITHMIC STEP SIZE DISTRIBUTION

If the step size is improperly chosen, in addition to losing accuracy, it would also result in numerical artifacts. For instance, the power of four wave mixing (FWM) products are usually overestimated by a constant step size method, since, FWM is a resonance effect. This can be avoided by using a logarithmic distribution for step sizes to keep the spurious FWM components below a certain level [14]. In this technique, for a fiber span of length L and loss coefficient α, the step size of the nth step is chosen as,

$$h_n = -\frac{1}{2\alpha} \ln\left(\frac{1-n\sigma}{1-(n-1)\sigma}\right) \tag{33}$$

where $\sigma = \dfrac{1-e^{-2\alpha L}}{K}$ and K is the number of steps per fiber span.

4. WALK-OFF METHOD

In a system which is dominated by dispersive effects, the step size is usually determined by the largest group velocity difference between the channels. In such cases, the step size is chosen so that, it is smaller than a characteristic walk-off length. The step size is given by $h = \dfrac{C}{\lambda_2 D_2 - \lambda_1 D_1}$ where λ_1 and λ_2 correspond to the smallest and the largest wavelengths in the system, D_1 and D_2 are the corresponding dispersions and C is a constant, which varies from system to system. This can be useful for single channel and multi channel systems.

5. LOCAL ERROR METHOD

In this method, the step size is adaptively controlled depending on the local error in each step. Starting from a specified z, the solution is evaluated for a step size 2h and then for a step size h, which are referred to as the coarse and fine solution respectively. A relative local error which is the ratio of relative error in the coarse solution with respect to the fine solution is estimated for the step, and depending on its numerical value, the step size is modified further. The computation involved is more laborious than a symmetric split step method, but this method yields a higher order solution, and a measure of the relative local error. This method works, irrespective of the amount of dispersive and nonlinear contribution [13].

The accuracy of a particular calculation, irrespective of the method used, can be estimated by calculating some conserved quantities such as the pulse energy (in the absence of absorption) along the length of the fiber. The energy of the pulse should be confined within the time window necessitating the choice of a sufficiently broad time window. Typically, window size is taken to be 10-20 times the pulse width.

Chapter 6

PROPAGATION CHARACTERISTICS OF PULSES IN DIFFERENT REGIMES OF OPERATION

6.1. PROPAGATION CHARACTERISTICS OF A GAUSSIAN PULSE

A detailed analysis of the propagation characteristics of a Gaussian pulse can be found in [5]. The salient features of the Gaussian pulse propagation are summarized below. The NLSE becomes completely analytically solvable, if the dispersion and nonlinear contributions are considered independently. The essential features in each of these cases in the time and frequency domain are discussed below.

6.1.1. Effect of Group Velocity Dispersion

Consider a Gaussian pulse with normalized power, for which the incident field at z = 0 is given by

$$A(0,T) = \exp\left(-\frac{T^2}{2T_0^2}\right) \tag{34}$$

where T_0 is the half width (at 1/e-intensity point). When this pulse propagates through a fiber in the dispersive regime, and when the terms contributing to the third order dispersion, self steepening and vibrational response effects are neglected, the NLSE reduces to a linear equation and can be integrated analytically. It is found that, after propagation,

1. The pulse maintains its shape but its width increases and the extent of broadening is decided by the propagation distance. For a given fiber length, short pulses broaden more because of the smaller dispersion length. It can be proved analytically that, at z= L_D, the pulse broadens by a factor of $\sqrt{2}$.
2. Though the incident pulse is unchirped (with no phase modulation) the transmitted pulse becomes chirped, which implies that its phase varies across the pulse. It is found that, the instantaneous frequency varies linearly across the pulse from the central frequency ω_0, and hence, the induced chirp is linear. This chirp depends on the sign of β_2. In the normal dispersion regime ($\beta_2 > 0$), the frequency difference from the central frequency is found to be negative at the leading edge (T<0) and increases linearly across the pulse; the opposite happens in the case of anomalous dispersion regime ($\beta_2 < 0$). The dispersion induced frequency chirp helps in the broadening of the pulse since different parts of the pulse evolve with slightly different frequencies and hence propagate at different speeds along the fiber.
3. For an initially unchirped Gaussian pulse, the dispersion-induced broadening of the pulse does not depend on the sign of the GVD parameter β_2. Thus, for a given value of the dispersion length, the pulse broadens by the same amount in the normal and anomalous dispersion regimes of the fiber.
4. Even when the initial Gaussian pulse is chirped, it maintains the shape on propagation. The broadening in this case would depend on the relative sign of the GVD parameter β_2 and the chirp parameter C. A Gaussian pulse broadens monotonically with z if $\beta_2 C > 0$, but it goes through an initial narrowing stage when $\beta_2 C < 0$.

6.1.2. Effect of Higher-Order Dispersion

If the wavelength of the pulse coincides with the zero-dispersion wavelength λ_D, then $\beta_2 = 0$ and hence, β_3 provides the dominant contribution to the GVD effects. For ultrashort pulses with widths $T_0 < 0.1$ ps, it is often necessary to include the β_3 term even when $\beta_2 \neq 0$. The dispersion length associated with higher-order dispersion is defined as

$$L'_D = \frac{T_0^3}{|\beta_3|} \tag{35}$$

The higher order dispersion effects play a significant role only if $L'_D \leq L_D$. The effects of the higher dispersion on the pulse shape as it propagates along the fiber can be summarized as follows:

1. Higher order dispersion distorts the pulse shape and the pulse becomes oscillatory near one of its edges.
2. For $\beta_3 > 0$, the oscillations appear near the trailing edge, while for $\beta_3 < 0$, the oscillations appear near the leading edge.
3. For $\beta_2 = 0$, oscillations are deep with intensity dropping to zero between the successive oscillations.
4. For $L'_D = L_D$, oscillations disappear, and the pulse has a long tail on the trailing edge.

6.1.3. Effect of Self Phase Modulation

The condition for invoking only nonlinear effects in the absence of dispersive effects is by making $L_D \gg L_{NL} \leq L$, where L is the fiber length. Since the nonlinear refractive index depends on the intensity, the phase of the emerging pulse varies according to the intensity of the pulse. It is seen that,

1. The phase-shift ϕ_{NL} of the output pulse is intensity dependent while the pulse shape in the time domain remains the same.
2. The non-linear phase-shift ϕ_{NL} is a function of propagation distance and increases linearly with the distance z. The phase-shift is maximum at T=0 (at the pulse center). The time dependence of ϕ_{NL} gives rise to the SPM-induced spectral broadening of the pulse over its initial width at z=0. Since the intensity, and hence the phase, changes across the pulse, the instantaneous frequency also changes across the pulse, in proportion to the temporal changes in the pulse, leading to a frequency chirp. When the pulse intensity reaches the maximum (at the center of the pulse), the

frequency chirp is zero. As the pulse intensity decreases in the leading edge of the pulse, the frequency chirp becomes positive, resulting in a blue shift. The intensity increases in the trailing edge of the pulse and hence is red shifted. Also, the chirp induced by SPM increases with the propagated distance and hence, new frequency components are continuously generated as the pulse propagates down the fiber.
3. Spectral broadening also depends on the shape and initial chirp C of the pulse. It is seen that a positive chirp increases the number of peaks in the spectrum and opposite occurs in the case of negative chirp. Since SPM-induced frequency chirp is linear and positive, it adds with the initial chirp for C > 0, resulting in an enhanced oscillatory structure. The opposite occurs in the case of negative chirp.

6.1.4. Combined Effect of Group Velocity Dispersion and Self Phase Modulation

When the dispersive and nonlinear lengths become identical, both GVD and SPM play equally important roles in the pulse evolution. Analytical solution is not possible for this case and the NLSE has to be solved numerically. The salient features for Gaussian pulse propagation, in a fiber such that the dispersion and nonlinear lengths are identical, are summarized below.

1. In the normal dispersion regime ($\beta_2 > 0$) the pulse broadens much more rapidly than that due to GVD alone. It is due to the fact that SPM generates new frequency components, which are red-shifted near the leading edge and blue-shifted near the trailing edge. In the normal dispersion regime, red components travel faster than the blue components, and hence, it helps to enhance the broadening of the pulse than expected from the GVD alone. Since the pulse shape changes in the time domain, the SPM induced broadening in the spectral domain becomes slightly lesser.
2. If the pulse is propagating in the anomalous dispersion regime ($\beta_2 < 0$), it first broadens at a rate much lower than that in the previous case and after a particular propagation distance, it reaches a steady state, beyond which there is no more broadening. In the spectral domain, the pulse shows a narrowing rather than broadening. This is attributed to the fact that SPM induces a positive chirp while GVD induces a negative chirp in the anomalous dispersion region. The two chirp contributions nearly cancel

each other along the central portion of the Gaussian pulse when $L_D = L_{NL}$. Pulse shape adjusts itself during propagation to make such cancellation as complete as possible. Thus, GVD and SPM cooperate with each other to maintain a chirp-free pulse, which is called a soliton. But the situation considered above does not correspond to ideal soliton evolution because the pulse initially broadens. But if a hyperbolic secant pulse is considered to start with, both the pulse shape and pulse spectrum remain unchanged.

The propagation characteristics discussed above are not universally true for all pulse shapes. They may drastically differ from the above depending on the type of the input pulse. To illustrate this, the propagation characteristics of an optical beat signal are discussed in the next section.

6.2. PROPAGATION CHARACTERISTICS OF A BEAT SIGNAL

In the previous sections, physical processes governing optical pulse propagation in fibers were summarized for a Gaussian pulse. In this section, the propagation of an optical beat signal through a fiber is discussed in detail, with the numerical results. The amplitude of the beat signal or the double side-band suppressed-carrier signal can be mathematically written as

$$A(0,T) = A_0 \cos\left(\frac{\pi T}{T_0}\right) \tag{36}$$

where T_0 is the beat period that is related inversely to the repetition rate of the beat signal. In time domain, this is a continuous-wave (CW) signal with sinusoidal modulation determined by the spectral difference between its two components. The beat signal can be generated by combining the outputs of two CW lasers with very narrow linewidth. The wavelength difference between the two lasers determines the frequency or repetition rate of the beat signal. The carrier wavelength of the beat signal will be the average of the wavelengths of the two lasers used to generate the beat signal. Since there is a π-phase difference between the adjacent lobes due to the sinusoidal nature of the beat signal, the nonlinear interaction among the adjacent pulses is very small and hence it maintains ultra-stable mark-to-space or period-to-duration ratio [15,16]. In addition, we can change the repetition rate of the pulses by just changing the wavelength difference between the two lasers. In the following sub-sections, we will consider the propagation of the beat signal at 10 GHz frequency through an optical fiber. For the beat signal input, the NLSE has to be solved numerically, using the SSFT technique discussed in section 5. The GVD parameter of the SMF is chosen to be -27.9 ps^2/km. The nonlinear refractive index n_2 and effective core area A_{eff} for these silica fibers are taken to be 3.2×10^{-20} m^2/W and 50 μm^2 respectively which corresponds to a nonlinear parameter (γ) of 2.6 W^{-1}km^{-1} and the loss is taken to be 0.2 dB/km. The propagation characteristics in different regimes are discussed in detail below.

6.2.1. Effect of Group Velocity Dispersion

The propagation of a 10 GHz ($T_0 = 100$ ps) beat signal at $\lambda_0 = 1549.72$ nm with 0.5 W of peak power through an optical fiber is considered, in the presence of only dispersion. In this case, only terms due to loss and GVD are included in equation (22). The values of β_2 and α corresponding to an SMF are used and the length of the fiber is chosen to be 4.2 km. Figure 2 shows the time domain, spectral domain, phase and chirp plots at the output. The FWHM of the input and output pulse are nearly the same as observed from Figure 2(a). The reduction in the peak power of the beat signal is entirely due to the losses in the fiber. The same argument holds for Figure 2(b) in which the input and output spectral domain plots are shown. Thus dispersion alone does not affect both the time and spectral domain of the beat signal at 10 GHz frequency. This is in contrast to the discussion in Section 6.1.1, where it is argued that dispersion broadens the time width and does not affect the frequency width of the pulse. Such an argument is true for the Gaussian and secant hyperbolic pulses, which contain a number of frequencies. In the case of a beat signal, since there are only two isolated frequencies involved, it is unaffected by dispersion. From the phase plot shown in Figure 2(c), it is obvious that the phase is constant across the entire pulse (indicating that dispersion does not change the phase) and adjacent lobes are π out of phase. This kind of phase implies that the chirp acquired by the beat signal is zero as seen in Figure 2(d). The vertical spikes are due to a phase change of π for each lobe. Hence, the beat signal can not be chirped with a purely dispersive element.

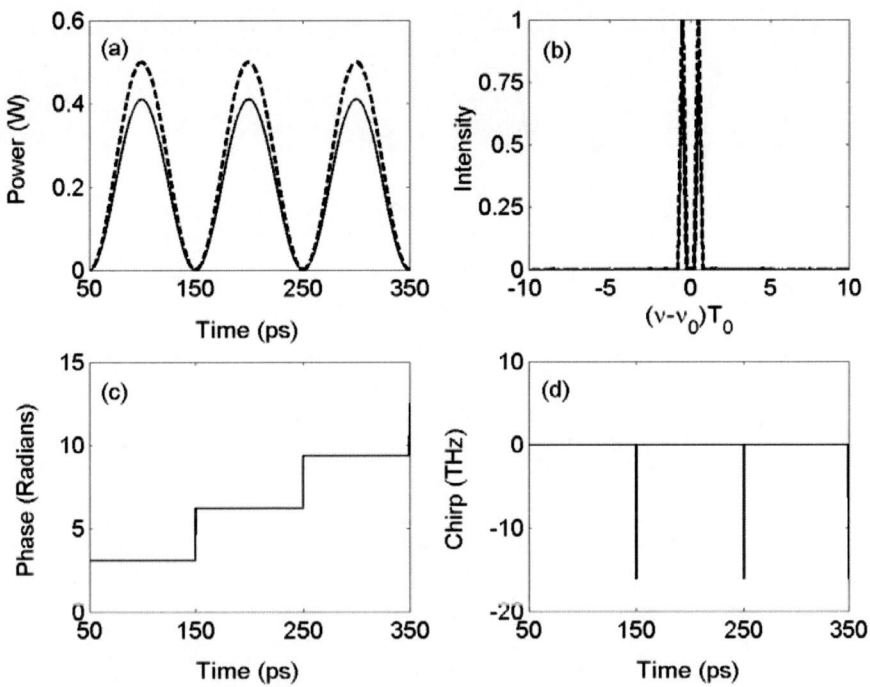

Figure 2. The evolution of 10 GHz beat signal through 4.2 km SMF including only dispersion (a) input (dashed line) and output (solid line) time domain, (b) corresponding spectral domain, (c) phase and (d) chirp acquired by the output pulse.

6.2.2. Effect of Self Phase Modulation

The propagation of 10 GHz beat signal with 0.5 W of peak power through the same length of the fiber is now considered, assuming that only nonlinearity and loss make a contribution to the pulse propagation. In this case, the nonlinearity changes the shape of the beat signal as expected. The reduction in the peak power is due to the absorption in the fiber. Comparing Figure 2(a) and Figure 3(a), it can be seen that there is no difference between the time domain shapes when dispersion and nonlinearity are considered independently. However, there is a broadening in the spectral domain of the beat signal due to nonlinearity, shown in Figure 3(b), similar to the result in Section 6.1.2. It can be seen from this figure that the broadened spectrum spreads from −4 to +4, which is equivalent to 80 GHz when the beat period T_0 is 100 ps.

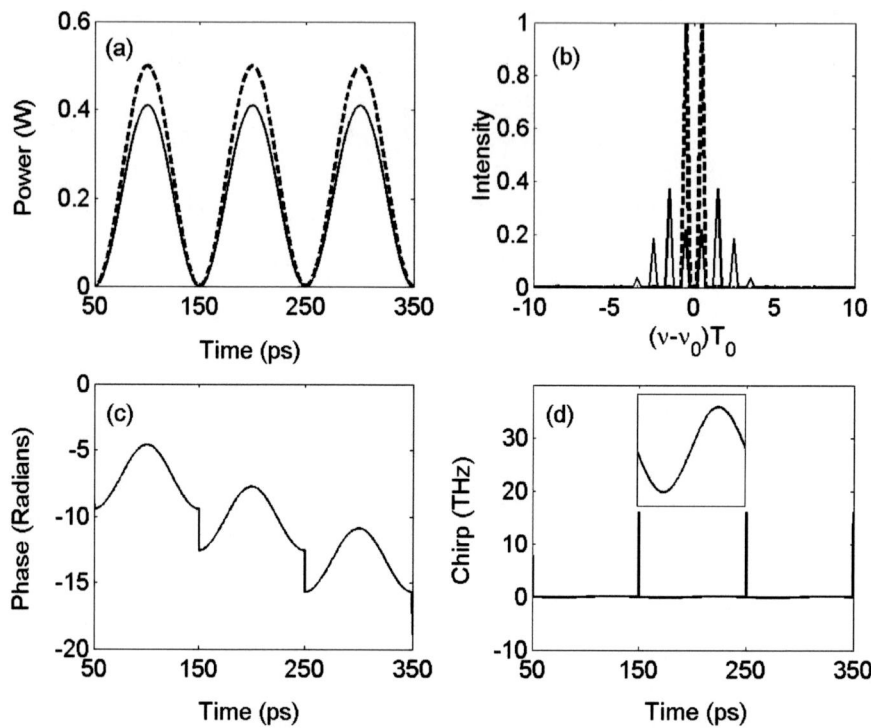

Figure 3. The evolution of 10 GHz beat signal through 4.2 km SMF including only SPM (a) input (dashed line) and output (solid line) time domain, (b) corresponding spectral domain, (c) phase and (d) chirp acquired by the output pulse (inset: expanded chirp across the central region of the pulse)

Figure 3(c) shows the phase as a function of time in which it is seen that the phase has the same shape as that of the beat signal. Since there is a variation in the phase with time, the pulse acquires a chirp that is linear and positive across its large central region as depicted in Figure 3(d). One more noteworthy feature is the oscillatory structure observed in the broadened spectrum in Figure 3(b). This behavior can be explained with the help of the theory elaborated in [5] for the Gaussian pulses. Inset in Figure 3(d) shows the expanded part of the chirp across one lobe of the beat signal. It is seen from the inset that the same value of the chirp occurs at two instants of time, indicating that the pulse has the same frequency at these two time instants. At these two points, the waves have the same instantaneous frequency but different phases (Figure 3(c)) and hence can interfere constructively or destructively depending on their relative phase difference. This leads to the multiple frequencies, seen as oscillations in Figure 3(b).

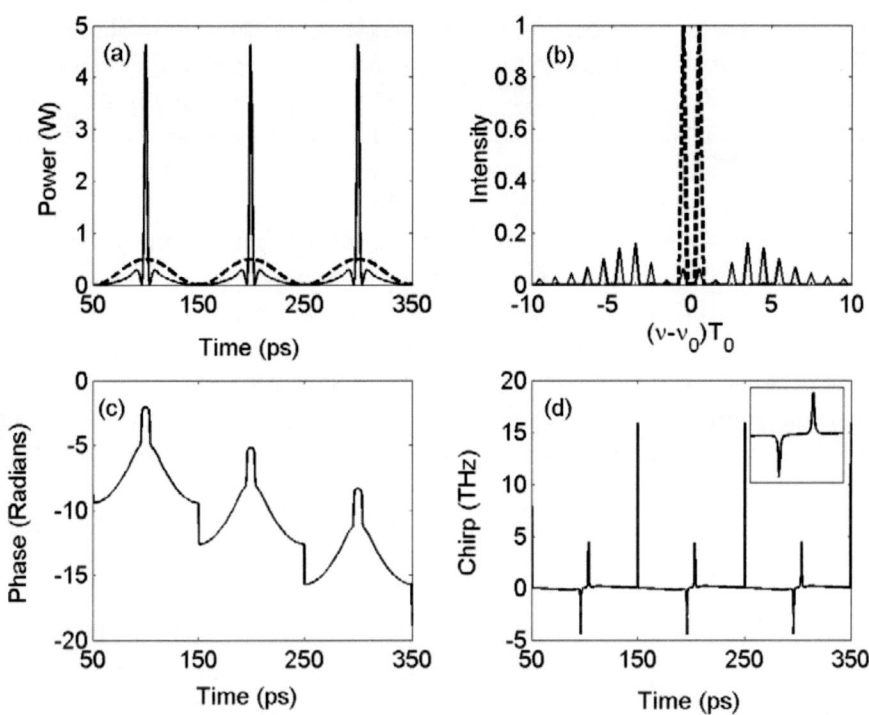

Figure 4. The evolution of 10 GHz beat signal through 4.2 km SMF including both dispersion and nonlinearity (a) input (dashed line) and output (solid line) time domain, (b) corresponding spectral domain, (c) phase and (d) chirp acquired by the output pulse (inset shows the expanded chirp across the central region of the pulse)

6.2.3. Effect of GVD and SPM

In this section, the effect of both anomalous dispersion and nonlinearity is considered. All the fiber parameters are the same as assumed in the previous cases. In the case of 10 GHz beat signal propagation, initially only nonlinearity acts and generates new frequency components thereby causing dispersion to act. SPM gives a positive chirp and anomalous dispersion introduces a negative chirp. Since these two have opposite effects, they may cancel each other leading to an output pulse that is narrower than the input pulse. At optimum compression length, the phase is nearly constant across the central region of the pulse.

In Figure 4(a), the output pulse at the end of 4.2 km SMF is shown in the time domain, and this length corresponds to optimum compression of this beat signal.

The output pulse has a peak power of nearly 4.6W with a FWHM of 2 ps indicating a compression factor (defined as the ratio of input pulse width to the output compressed pulse width [17]) of 25. Due to the narrowing of the pulse in the time domain, there is a broadening in the spectral domain as observed in Figure 4(b). The phase is almost flat across the compressed pulse width as shown in Figure 4(c) resulting in zero chirp across the FWHM of the pulse (Figure 4(d)). Inset in Figure 4(d) shows the expanded chirp for the central region of the pulse, where the zero value and the flat nature of the chirp can be clearly seen. It can also be observed that, the compressed pulse train contains broad pedestals in the time domain, due to which there is a central dip in the output spectrum. The origin of the pedestals can be explained by the fact that SPM-induced chirp is linear only across the central region of the pulse; thus anomalous dispersion cancels the chirp associated only with the central part of the pulse. In effect, anomalous GVD compresses only the central part of the pulse and hence energy in the pulse wings remains uncompressed and appears in the form of pedestals. Thus, a 10 GHz beat signal can be compressed in the time domain using an SMF of appropriate length.

Chapter 7

SPECIFIC APPLICATIONS OF THIRD ORDER NONLINEARITY

7.1. MULTI WAVELENGTH OPTICAL SOURCE FOR WDM APPLICATIONS

In 1980s, optical fibers were used for data transmission in a single channel mode. A common approach was to employ time-division multiplexing (TDM) of the data pulses. However, it is not possible to use electronic TDM for data rates far greater than 10 Gb/s with the present technology. Since the bandwidth of an optical fiber can support data rates larger than gigahertz (GHz) that can be realized by TDM, use of other multiplexing schemes was explored. Optical channel multiplexing can be done either in time domain or in wavelength domain. This leads to optical time division multiplexing (OTDM) and wavelength division multiplexing (WDM) respectively. In OTDM, several optical signals modulated at the bit-rate B using the same carrier frequency are multiplexed optically to form a composite signal at the bit rate N × B, where N is the number of channels. This requires a laser source that can be modulated to yield pulses at a repetition rate equal to the bit-rate B and with pulse width $< (NB)^{-1}$ to ensure that the pulses fit in the allocated time slot. Even though the available spectrum is efficiently used in OTDM, the complex multiplexing, de-multiplexing schemes and cost factors make the OTDM more difficult to implement. In addition, chromatic dispersion is the main limiting factor in OTDM.

7.1.1. Wavelength Division Multiplexed Sources

In WDM, multiple signals at slightly different wavelengths are simultaneously transmitted through a single fiber [17,18]. If in an N-channel WDM system, each channel contains a TDM data at a bit rate B, and then the total bit-rate capacity of the system becomes N × B. In other words, to achieve data rates equal to OTDM, the number of optical carriers is increased N times leading to a reduction in the bit rate by a factor of N. The lower data rate reduces the effect of dispersion. The increasing demand on high-bit-rate communication systems requires more and more number of channels to be multiplexed. In such systems, the inter-channel spacing is reduced to 100/50 GHz and these systems are called as dense wavelength division multiplexed (DWDM) systems [19]. WDM has gained popularity mainly due to the negligible deterioration from chromatic dispersion effects in this scheme. The main advantages of WDM are its higher capacity, easy upgradability and lower cost. An important aspect that puts a limit on the number of channels in a WDM system is the stability of distributed feedback (DFB) semiconductor laser diodes, which are widely used as sources in WDM transmission system. When many WDM channels are present, it becomes quite expensive to provide a wavelength-stabilized source for each channel. Some of the alternatives to overcome this problem are provided in schemes such as

i. multi-wavelength DFB laser diode arrays [20-22]
ii. multi-wavelength birefringent-cavity mode-locked fiber ring lasers [23,24]
iii. multi-frequency laser with waveguide grating router as the intracavity filter element [25-27]
iv. WDM source based on spectral slicing of broadband spectrum of incoherent sources such as light emitting diode (LED) and superluminescent diode (SLD) or broadband amplified spontaneous emission (ASE) of an EDFA [28-31], and
v. multi-wavelength pulsed source generated by spectral slicing of broadband spectrum associated with a single pulsed source. Such broadband spectra are available with the supercontinuum (SC) light generated in optical fibers, femtosecond pulsed sources and pulses generated by nonlinear soliton compression [32-46]. SC or broadband spectrum generated by pulse propagation in an optical fiber has attracted considerable attention since the last few years.

7.1.2. Multiwavelength Source Using a Beat Signal

A parametric pulse source generated by optically beating two CW sources operated at slightly different wavelengths is an attractive way of generating the optical pulses at high repetition rate without the use of mode-locking. These high rep-rate pulses can be further spectrally enriched through nonlinear mixing process [47-51]. In this scheme to generate a broad spectral content, an optical beat signal with appropriate time domain characteristics is amplified and allowed to propagate through a low dispersion fiber, so as to enhance the nonlinear processes. Depending on the time domain characteristics of the input beat signal, a pre-chirp is introduced before its propagation through the nonlinear fiber. This can be done by allowing it to propagate through an appropriate length of a standard single mode fiber. Figure 5 shows the schematic representation of the time and wavelength domain of the output from such a scheme, by solving the NLSE using the SSFT technique discussed earlier. The third order dispersion and Kerr nonlinear operators are included and appropriate step sizes are chosen. A spectral enrichment of the extent of about 3.2 nm is obtained with this scheme.

The output from this source can be filtered using a demultiplexer with an interchannel spacing of 100 GHz. All the filtered channels in this case, would be synchronized in time with a data rate of 10 Gbps and this source would be suitable for synchronous data transfer at 16×10 Gbps in applications such as Bit Parallel-WDM. If the wavelength separation between the two sources is 0.8 nm, it can also be used for asynchronous data transfer, by slicing the spectral content using a WDM demultiplexer of 100 GHz spacing. Each of the spectrally sliced multiple channels will be available as CW, which may be further externally modulated for transmission purposes.

Figure 6 shows the schematic of the experimental set up for this scheme. The beat signal is generated by mixing the outputs of two narrow line width DFB laser sources, which have a wavelength separation of 0.08 nm (corresponding to a frequency separation of 10 GHz), around 1550 nm and is amplified to an average value of around 100 mW using a bidirectionally pumped EDFA [52-54].

The spectrally enriched output observed experimentally for a 10 GHz beat signal is shown in Figure 7. The broadband at the output indicates the possibility to get 6 channels at 100 GHz separation. The output spectrum can be demultiplexed, as discussed earlier, and could be used as a source for DWDM communication systems. The scheme of beat signal generation has the flexibility of easy up-gradation to higher bit-rates without the use of cumbersome RF electronic circuitry as in the case of harmonic mode-locking. In passive mode-locking, it is quite difficult to achieve pulses of repetition rate much greater than

10 GHz whereas active mode-locking suffers from the disadvantages of pulse break-up and environmental instabilities. The other advantage with the beat signal is that the adjacent pulses have π-phase difference between them, thus reducing the interaction among the pulses leading to a stable period-to-duration ratio.

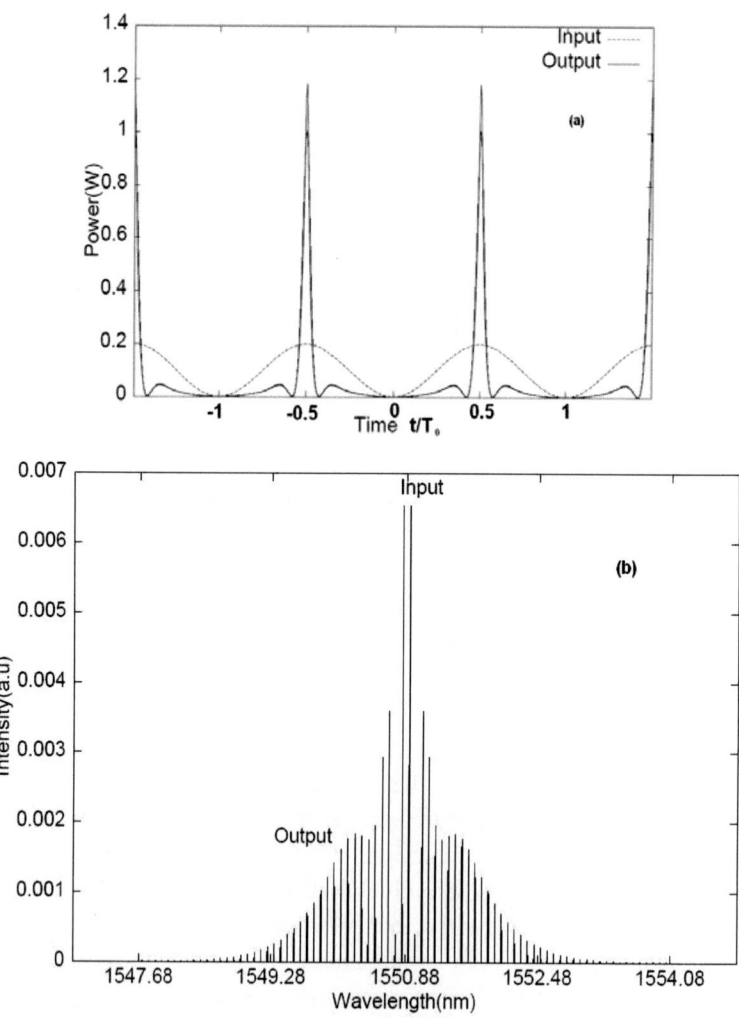

Figure 5. (a) Amplified beat signal at 10 GHz and the shaped output pulse after propagation through 7 km of SMF and 1 km DSF in the time domain (b) Input and output in the spectral domain. The center wavelength of the pulse is 1550.88 nm

Specific Applications of Third Order Nonlinearity   37

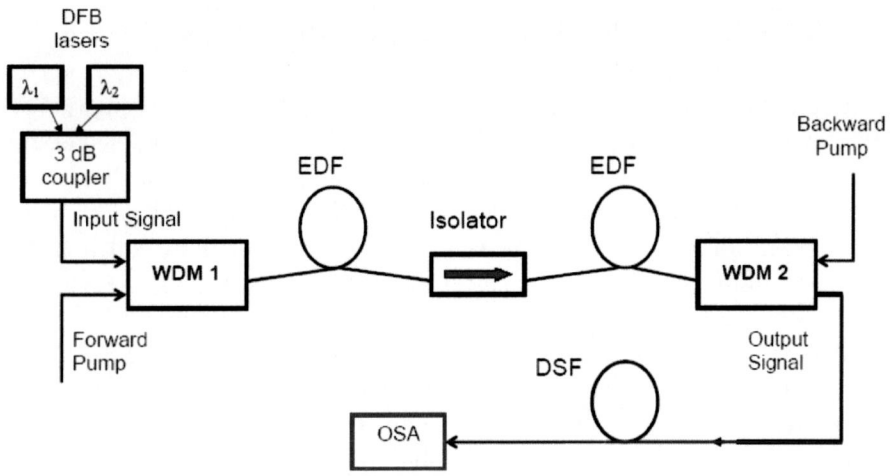

Figure 6. Schematic of the experimental set up to generate multiple wavelengths, starting from a beat signal.

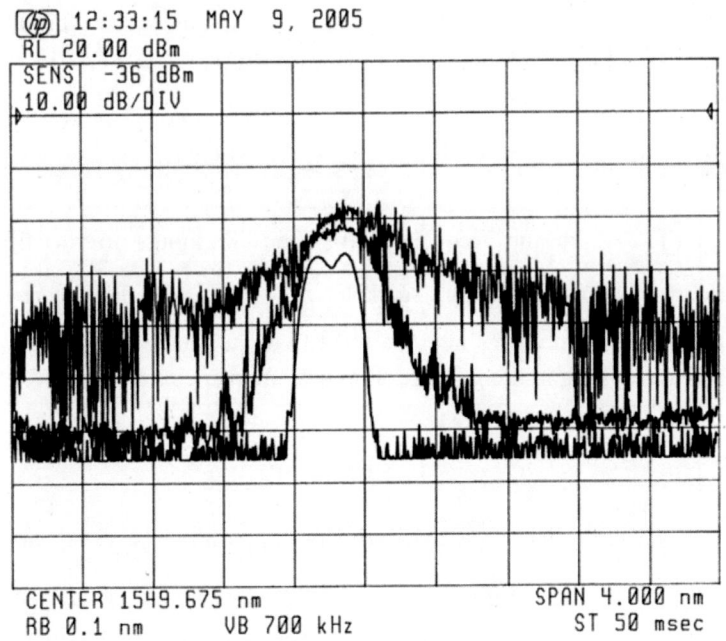

Figure 7. Input 10 GHz beat signal and shaped spectrum through DSF at lower and higher (75 mW) powers (in 4 nm span).

7.2. FOUR WAVE MIXING AND SOME APPLICATIONS

One of the prominent nonlinear effects in fibers is four wave mixing (FWM). FWM is the process by which, electromagnetic fields of different frequencies, propagating simultaneously in a nonlinear medium, interact through the third order nonlinear optical susceptibility of the medium, resulting in the generation of new frequencies. FWM in single mode fibers was first reported by Hill et al [55]. Since it can influence the multi channel transmission systems significantly, this phenomenon has been studied extensively in different types of fibers [56-62]. For FWM to be efficient, the phase matching condition has to be satisfied [63,5]. It is often indicated that, in the case of fibers, the FWM efficiency at any wavelength is degraded due to the chromatic dispersion of the fiber at that wavelength [64]. Hence, the use of dispersion shifted fiber (DSF) and choice of the signal wavelengths close to or at the zero dispersion wavelength are suggested as the best options to achieve phase matching and hence higher FWM efficiencies. However, at high signal powers, the non-linearity in the fiber also contributes to the phase mismatch through self-and cross-phase modulation. Therefore, it is the total phase mismatch due to the dispersion and the non linearity of the fiber which decides the FWM efficiency in optical fibers [65,66].

7.2.1. Theory

Let f_i, f_j, f_k be the frequencies of the *CW* inputs to the fiber, with propagation constants k_i, k_j, k_k and let A_i, A_j, A_k be their corresponding amplitudes. These input waves interact in the fiber due to the third order nonlinear susceptibility, resulting in the generation of new frequencies at

$$f_{ijk} = f_i + f_j - f_k \tag{37}$$

(subscripts *i,j,k* can select 1,2,3, $j \neq k$) with amplitude, A_{ijk} and propagation constant, k_{ijk}. If two of the input frequencies are the same, it is referred to as the partially degenerate FWM. If all the frequencies are equal, it is totally degenerate FWM. Assuming that the pump waves are not depleted due to the generation of the FWM products, the coupled differential equations for the propagating amplitudes, including the contributions to phase mismatch due to XPM and SPM can be written as [66]

$$\frac{dA_i}{dz} = -\frac{1}{2}\alpha A_i + 2i\gamma\left(|A_i|^2 + 2|A_j|^2 + 2|A_k|^2\right)A_i \quad (38)$$

$$\frac{dA_j}{dz} = -\frac{1}{2}\alpha A_j + 2i\gamma\left(2|A_i|^2 + |A_j|^2 + 2|A_k|^2\right)A_j \quad (39)$$

$$\frac{dA_k}{dz} = -\frac{1}{2}\alpha A_k + 2i\gamma\left(2|A_i|^2 + 2|A_j|^2 + |A_k|^2\right)A_k \quad (40)$$

$$\frac{dA_{ijk}}{dz} = -\frac{1}{2}\alpha A_{ijk} + 2i\gamma\left(|A_i|^2 + |A_j|^2 + |A_k|^2\right)A_{ijk} + \frac{1}{3}Di\gamma A_i A_j A_k^* e^{i\Delta kz} \quad (41)$$

D is the degeneracy factor (D = 2, 3, 6 for totally degenerate, partially degenerate and non degenerate case respectively), and Δk is the total phase mismatch which has contributions from dispersion (Δk_l) and nonlinearities (Δk_{nl}).

$$\Delta k = \Delta k_l + \Delta k_{nl} \quad (42)$$

The nonlinear contribution to the phase mismatch can be analytically derived as [66],

$$\Delta k_{nl} = \gamma\left(P_i + P_j - P_k\right)\left(\frac{1-e^{-\alpha L_{eff}}}{\alpha L_{eff}}\right) \quad (43)$$

where

$$L_{eff} = \frac{1-e^{-\alpha L}}{\alpha} \quad (44)$$

The dispersive phase mismatch can be calculated as

$$\Delta k_l = k_{ijk} - (k_i + k_j - k_k) \tag{45}$$

Expanding the propagation constants in a Taylor series about $\omega_k = 2\pi f_k$ and retaining terms up to third order in $\omega - \omega_k$ the linear phase mismatch at a given wavelength can be written in terms of the dispersion and dispersion slope. However, if the wavelength separation between the mixing wavelengths is large, this approach has to be modified [67]. The mixing efficiency for the Stokes and anti Stokes components needs to be calculated differently, since their separation and hence the dispersion is significantly different. For instance, in the case of partially degenerate four wave mixing (mixing of two wavelengths) resulting in Stokes and antiStokes components as shown in Figure 8, the dispersive phase shifts are given by

$$\Delta k_{l3} = \left\{ \frac{4\pi\lambda_3^2}{c}(f_1 - f_2)^2 D \right\}_c - \left\{ \frac{6\pi\lambda_3^2}{c^2}(f_1 - f_2)^3 \left[2\lambda_3 D_c + \lambda_3^2 \frac{dD_c}{d\lambda} \right] \right\} \tag{46}$$

$$\Delta k_{l4} = \left\{ \frac{4\pi\lambda_4^2}{c}(f_1 - f_2)^2 D_c \right\} + \left\{ \frac{6\pi\lambda_4^2}{c^2}(f_1 - f_2)^3 \left[2\lambda_4 D_c + \lambda_4^2 \frac{dD_c}{d\lambda} \right] \right\} \tag{47}$$

where D_c and $\frac{dD_c}{d\lambda}$ are evaluated at the wavelengths at which the phase mismatch is calculated. It should be noted that, the dispersive phase shift is positive in the normal dispersion region and negative in the anomalous dispersion region. Since the nonlinear phase shift is always positive, it is easier to phase match $(\Delta k = 0)$ in the anomalous dispersion region than at the zero dispersion wavelength. The power in the generated FWM products, and hence the efficiency of generation of the sidebands, can be calculated using

$$P_{ijk} = \frac{\eta}{9} D^2 \gamma^2 P_i P_j P_k e^{-\alpha L} \left\{ \frac{[1 - e^{-\alpha L}]^2}{\alpha^2} \right\} \tag{48}$$

with η being the FWM efficiency, defined as

$$\eta = \frac{\alpha^2}{\alpha^2 + \Delta k^2}\left\{1 + \frac{4e^{-\alpha L}\sin^2(\Delta kL/2)}{(1-e^{-\alpha L})^2}\right\} \tag{49}$$

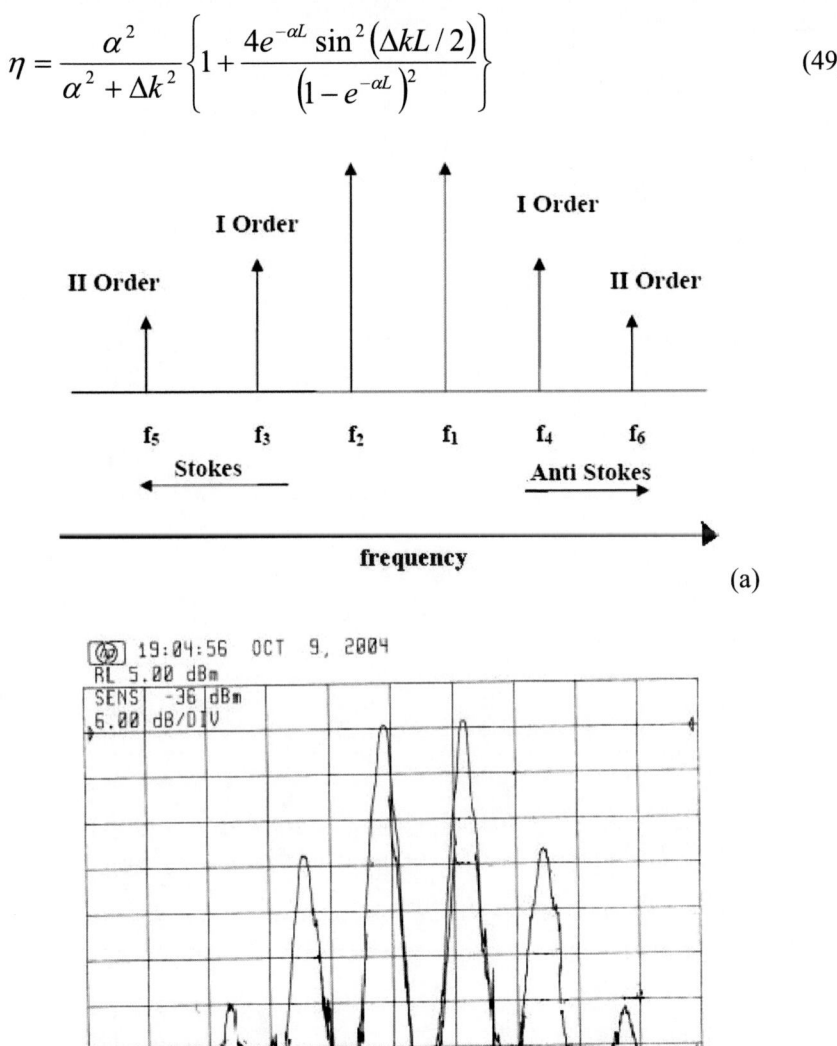

Figure 8(a): Pictorial representation of the generation of I and II Order Stokes and AntiStokes waves due to FWM (b) The FWM products observed experimentally from a low dispersion fiber of length 5 km.

The estimated values of efficiencies calculated from equations (46) and (47) for a fiber length of 5 km, and the corresponding experimentally observed values are shown in Figure 9.

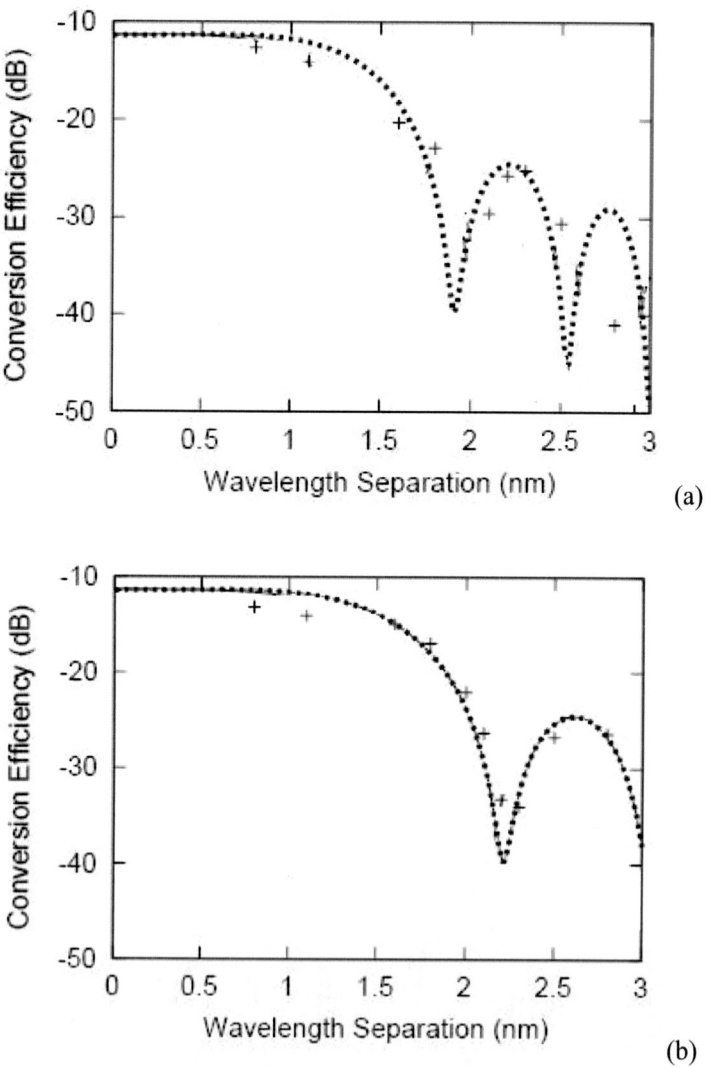

Figure 9 Variation of conversion efficiency with wavelength separation for I order (a) Stokes and (b) anti Stokes components for a DSF length of 5 km and pump power of 34 mW each. The discrete points show the corresponding experimental data. The DSF has its zero dispersion wavelength at 1544 nm and a dispersion slope of 0.072 $ps/km\text{-}nm^2$.

Figure 9 shows that the conversion efficiencies of the Stokes and antiStokes components are not identical for a given wavelength separation. Since the conversion efficiency is an oscillatory function of the phase mismatch, it has to be calculated correctly. The experimentally observed data is found to follow closely to the expected efficiencies.

Second order FWM products can be formed due to the mixing between more than one combination of frequencies. For instance, the second order Stokes frequency, f_5 is generated primarily due to the mixing of

1. f_1, f_2, and f_3 (non degenerate)
2. f_2 and f_3 (partially degenerate)
3. f_2 and f_4 (partially degenerate)

In addition to the above possibilities, there could be contributions due to the mixing of higher order frequencies. The individual linear phase mismatch due to each of the above contributing term is calculated by expanding the propagation constants at each frequency as a Taylor series about f_5. Retaining up to the third order term, the analytical expressions for the dispersive phase mismatches at the generated wavelength, λ_5 can be written as follows.

$$\Delta k_{l51} = \left\{\frac{4\pi\lambda_5^2}{c}(f_1 - f_2)^2 D_c\right\} - \left\{\frac{6\pi\lambda_5^2}{c^2}(f_1 - f_2)^3 \left[2\lambda_5 D_c + \lambda_5^2 \frac{dD_c}{d\lambda}\right]\right\}$$
$$+ \left\{\frac{32\pi\lambda_5^2}{c^3}(f_1 - f_2)^4 \left[\lambda_5^2 D_c + \lambda_5^3 \frac{dD_c}{d\lambda}\right]\right\} \quad (50)$$

$$\Delta k_{l52} = \left\{\frac{2\pi\lambda_5^2}{c}(f_1 - f_2)^2 D_c\right\} - \left\{\frac{2\pi\lambda_5^2}{c^2}(f_1 - f_2)^3 \left[2\lambda_5 D_c + \lambda_5^2 \frac{dD_c}{d\lambda}\right]\right\}$$
$$+ \left\{\frac{7\pi\lambda_5^2}{c^3}(f_1 - f_2)^4 \left[\lambda_5^2 D_c + \lambda_5^3 \frac{dD_c}{d\lambda}\right]\right\} \quad (51)$$

$$\Delta k_{l53} = \left\{\frac{8\pi\lambda_5^2}{c}(f_1 - f_2)^2 D_c\right\} - \left\{\frac{16\pi\lambda_5^2}{c^2}(f_1 - f_2)^3 \left[2\lambda_5 D_c + \lambda_5^2 \frac{dD_c}{d\lambda}\right]\right\}$$

$$+\left\{\frac{112\pi\lambda_5^2}{c^3}(f_1-f_2)^4\left[\lambda_5^2 D_c + \lambda_5^3 \frac{dD_c}{d\lambda}\right]\right\} \quad (52)$$

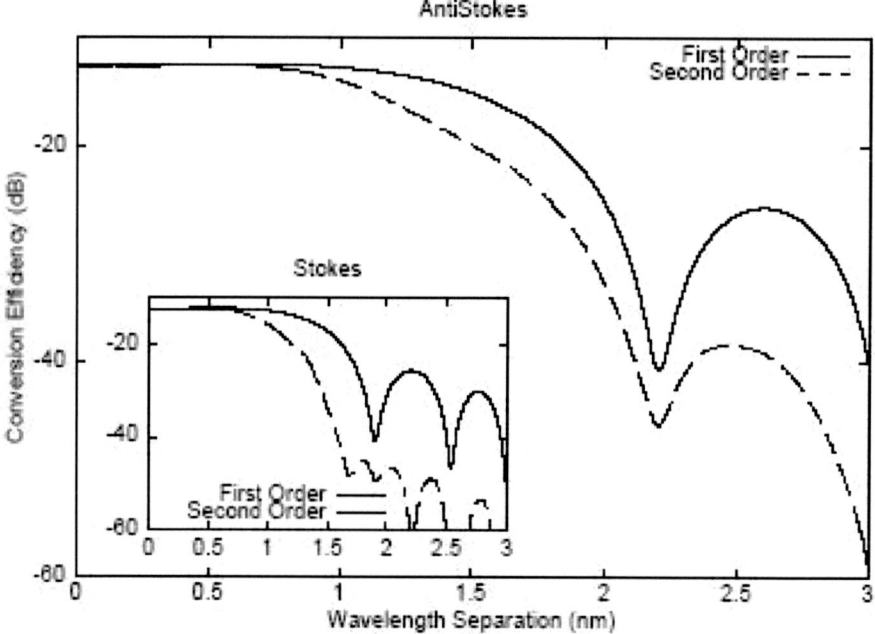

Figure 10. Variation of conversion efficiencies at the anti Stokes wavelength with wavelength separation between the mixing wavelengths, at individual power of 30 mW each and a DSF length of 5 km. The corresponding variation for the Stokes wavelength is shown in the inset.

Similar expressions can be derived for the antiStokes FWM products. While calculating the efficiencies of generation of the second order product at a particular frequency, eqn (51) has to be used, considering the interference between the generated frequencies since the contributing terms to generate a particular frequency need not be in phase. For instance, the efficiency of generation of the second order Stokes wavelength, which has three major contributions discussed above, can be written as

Specific Applications of Third Order Nonlinearity

$$P_5 = P_{51} + P_{52} + P_{53} + 2\sqrt{P_{51}P_{52}}\cos[(\Delta k_{51} - \Delta k_{52})L] + 2\sqrt{P_{52}P_{53}}\cos[(\Delta k_{52} - \Delta k_{53})L]$$

$$+ 2\sqrt{P_{51}P_{53}}\cos[(\Delta k_{51} - \Delta k_{53})L] \tag{53}$$

The variation of conversion efficiencies with wavelength separation between two mixing wavelengths in the case of degenerate four wave mixing, for the first and second order products using the equations above is shown in Figure 10.

The following are the inferences in the case of a degenerate four wave mixing model as seen from Figure 10.

1. The variation in the conversion efficiencies with wavelength separation for the Stokes and the anti Stokes wavelengths are not identical for both the first and second order FWM products since they are differently positioned with respect to the zero dispersion wavelength.
2. Nonlinear phase shifts contribute significantly to the phase matching mechanism, since the Stokes and antiStokes wavelengths, in spite of being in the anomalous dispersion region, remain phase matched for larger wavelength separation.
3. Perfect phase matching at a given power is possible only at those sideband frequencies where the phase shift due to dispersion is exactly compensated by the phase shift due to the nonlinearity. It is often implied that, in the case of phase matching due to nonlinearity, the frequency of the sidebands depends on the power of the input signals [5]. Instead it is worth noting that the frequency of the sidebands is always power independent, and is given by equation (37).
4. The second order products are as efficient as the corresponding first order products for smaller wavelength separations. At this power, the efficiency of generation of the second order Stokes and anti Stokes components are also the same, for a wavelength separation of 0.8 nm. This indicates that the generation of higher order FWM components with efficiencies as high as the first order components is theoretically feasible.
5. Multiwavelength generation is possible through FWM by starting with two CW frequencies and by carefully designing the fiber length, power, the position of pump wavelengths with respect to the zero dispersion wavelength and their wavelength separation.
6. The above theory is valid under the assumption that the powers in the generated first and second order sidebands are much less than the pump waves. This undepleted pump power approximation is not quite valid for

larger conversion efficiencies, and hence these results show only the expected trends. The exact powers in the sidebands can be estimated using the Split Step Fourier Transform method.

Optical communication is conventionally handled in the C band (1530-1560 nm), and fibers with zero dispersion wavelength shifted to this band are recommended in multi-wavelength systems. In WDM systems, FWM poses a serious problem where the introduction of different carrier wavelengths is limited by the mutual interaction between the channels. Several techniques are suggested to suppress the efficiency of the process, which include specialized modulation of the channels [68,69]. New techniques for the suppression of the four wave mixing induced distortion include nonzero dispersion fiber WDM systems [70], introduction of additional devices into the transmission system [71], reduction of channel power, and using unequally spaced carrier frequencies [72], and effective channel allocation to reduce in-band FWM crosstalk in DWDM transmission systems [73].

7.2.2. Some Applications of FWM

Four wave mixing provides an effective technological base for many fiber optic devices and applications. Some of them are discussed below.

One of the most promising devices in the photonic networks of future is a wavelength converter. A wavelength converter is a device which converts data from an input wavelength into another output wavelength. It thus helps to dynamically reconfigure WDM networks to improve the network blocking performance. Since electronic polarization, which is an instantaneous effect, is responsible for the FWM process, the wavelength converters based on FWM are not limited to the modulation format or to the bit rate of the systems. A strong pump wave with an appropriate wavelength shift and an optical fiber with enhanced nonlinearity are the essential components of a wavelength converter. Usually, a highly nonlinear dispersion shifted fiber is used to enhance the conversion bandwidth [74]. Specific designs of parametric amplifiers are available to minimize the cross talk, when more than one signal is launched [75-77]. It is also reported that, by using a loop configuration and appropriately selecting the fiber and the pump, the entire C band can be converted to S band and even L band [78].

The deployment of high speed optical communication networks and the use of non-conventional fibers for multiple purposes have made the measurement of chromatic dispersion and nonlinear coefficient extremely important. FWM serves as a tool to simultaneously and efficiently measure the nonlinear coefficient, zero dispersion wavelength and chromatic dispersion in dispersion shifted fibers. Two wavelengths are simultaneously launched into the fiber under test and the efficiency of the generated FWM products are monitored carefully while one of the wavelengths is continuously changed. The theory of degenerate FWM discussed above is used to fit to the observed results. Since the efficiency of the process depends on the nonlinear index coefficient and the relative position of the mixing wavelengths with respect to the zero dispersion wavelength, all these parameters can be extracted from the theoretical fit [79]. The spatial variation of zero dispersion wavelength along the length of the fiber can also be measured by using short pulses, which overlap only in a small portion of the fiber due to GVD induced walk- off [80]. This technique is non-destructive and is useful to obtain the dispersion map of long lengths of fibers. The effect of polarization mode dispersion also needs to be accounted, since the random nature of PMD results in fluctuations in the idler power generated through FWM, and hence would affect the mapping of the zero dispersion wavelength and chromatic dispersion. A detailed analysis of the same can be found in Q Lin et al [81].

Another novel application of four wave mixing is in all-optical signal reshaping of digital high-bit rate signals [82,83] and pulse compression [84]. This is possible by making use of the step-like all-optical transfer function for the higher order terms in four wave mixing. Noise compression on spaces occurs due to the characteristic dependence of FWM generated waves on input power. At higher pump powers, since the pump depletion leads to gain saturation, this results in noise compression on marks as well.

7.3. STIMULATED BRILLOUIN SCATTERING AND SOME APPLICATIONS

Brillouin scattering is a process resulting from the interaction between incident an optical field and the acoustic waves in the material [3]. The thermo-elastic motion of the molecules in the material leads to density fluctuations within the medium. These density fluctuations propagate through the medium with the velocity of sound and they are considered as acoustic waves. The periodic density modulation results in a refractive index modulation, and causes diffraction of the incident waves. Since the density modulation moves, and hence the refractive index grating moves with an acoustic velocity v_A relative to the incident waves, the scattered wave experiences a frequency shift due to Doppler effect. Unlike this *spontaneous* effect, if the incident light wave itself generates the acoustic wave through electrostriction, the effect is called *Stimulated* Brillouin Scattering (SBS), in which case, a large part of the power is transferred to the scattered wave [3]. The scattered wave in the backward direction is downshifted in frequency, and hence called the Stokes wave. Quantum mechanically, the phenomenon is described as the annihilation of a pump photon and the creation of a new photon with a downshifted frequency (Stokes) and a phonon. Thus the process can be interpreted as a parametric acousto-optic interaction between the pump photon, the Stokes photon and the acoustic waves.

7.3.1. Theory

The conservation of energy and momentum in the scattering process demands

$$\Omega_B = \omega_p \pm \omega_s \text{ and } k_A = k_p \pm k_s \tag{54}$$

where ω_p, ω_s and Ω_B are the frequencies; k_p, k_s and k_A are the wave vectors of the incident pump, the Stokes and the acoustic waves respectively. Eqn (54) is written for the extreme cases corresponding to the conditions where the pump and the acoustic waves propagate along the same direction (corresponding to the positive sign) or the pump and the acoustic waves move in opposite directions (corresponding to the negative sign). In general, the acoustic wave satisfies the standard dispersion relation,

$$\Omega_B = v_A |k_A| \approx 2 v_A k_p \sin(\theta/2), \tag{55}$$

where θ is the angle between the pump and Stokes fields. Eqn (55) shows that the frequency shift of the Stokes wave depends on the angle of scattering and is a maximum in the backward direction $(\theta = \pi)$, corresponding to the case where the acoustic waves and the pump wave move in opposite directions. It can also be seen that, there is no stimulated scattering in the forward direction, though spontaneous scattering can occur in the forward direction. In the case of SBS in single mode optical fibers, the relevant direction is $\theta = \pi$, since guidance is not possible in all other directions. Hence, for optical fibers, the Brillouin frequency shift can be written as

Specific Applications of Third Order Nonlinearity

$$v_B = \frac{2nv_A}{\lambda_p} \tag{56}$$

where n is the effective modal index for the pump wavelength, λ_p. Thus, the frequency shift depends on the wavelength of the pump, velocity of the acoustic wave and refractive index. For silica fibers the typical frequency shift at a pump wavelength of 1.55 μm is about 11.1 GHz. The spectral width of the Brillouin gain spectrum depends on the lifetime of the acoustic phonons. It is usually approximated as a Lorentzian function [3], for quasi monochromatic pump. The peak value of gain coefficient is given as [85,5]

$$g_B = \left(\frac{4\pi n^8 p_{12}^2}{c\lambda_p^3 \rho_0 v_B \Delta v_B} \right) \left(\frac{\Delta v_B}{\Delta v_B \otimes \Delta v_p} \right) \tag{57}$$

where p_{12} is the longitudinal elasto-optic coefficient, Δv_B is the full width at half maximum of the gain spectrum(which is the reciprocal of the phonon life time), Δv_P is the linewidth of the pump at wavelength λ_p, and ρ_0 is the material density. For silica fibers, with CW pump, the gain coefficient is estimated to be $5 \times 10^{-11} mW^{-1}$. For Lorentzian gain profile, the convolution in the denominator of eqn (57) can be calculated as $\Delta v_B + \Delta v_p$. The gain curve is reported to narrow with the increase in pump power [86]. In the case of temporally short pump pulses, the spectrum is correspondingly broad and the gain curve converges to a Gaussian function as a result of the superposition of a number of Lorentzian functions. The width of the gain spectrum in this case is decided by the duration of the pump pulse, and the gain spectrum gets narrowed when the pump pulse is shorter than the phonon life time. For pulses shorter than 5 ns, the gain spectrum becomes identical to the CW value [87].

7.3.2. Threshold for SBS

It is found that, as the input pump power increases, the throughput power from the fiber also increases. But when the pump power exceeds a certain distinct value, the throughput power is found to saturate. This indicates the onset of SBS.

Beyond this threshold, the Stokes power is also found to increase. Brillouin threshold is defined in different ways in the literature. Some of them are [88]:

1. The input pump power at which the Stokes power becomes equal to the input pump power.
2. The input pump power for which the Stokes power and the throughput power become equal.
3. The input pump power for which the backscattered Stokes power increases rapidly and the pump power saturates.
4. The input power for which the Stokes power is equal to 1% of the input pump power.

Following ITU recommendations, the threshold power is calculated for CW pump using the formula [88],

$$P_{th} \cong 19 \frac{KA_{eff}}{g_B L_{eff}} \tag{58}$$

where K is the polarization factor, which varies from 1 to 2 depending on the polarization state of the pump and the signal. Thus larger lengths of the fiber would indicate lower threshold powers. SBS threshold in communication systems is found to be in the range of 1-10 mW, due to this reason.

7.3.3. Experimental Determination of SBS Threshold

SBS threshold corresponding to any of the definitions listed above can be measured experimentally by launching pump with varying powers to the fiber under test and measuring the throughput and backscattered power. A typical experimental setup is as shown in Figure 11.

The optical power from a narrow line width DFB laser source is amplified with an Erbium doped fiber amplifier (EDFA) and is allowed to propagate through the fiber under test placed after a circulator. The throughput and reflected powers are measured using optical power meters. A variable optical attenuator (VOA) is included so as to control the power entering the fiber. As the input power increases, in addition to the increase in throughput power, the backscattered power is also found to increase linearly due to the Fresnel reflections from the far end. As the launched power approaches the threshold power, the backscattered power increases rapidly and beyond the threshold, the throughput power saturates and the input power is converted to Stokes power. The variation of transmitted power as a function of the input CW power for different lengths of a dispersion shifted fiber (DSF), observed experimentally is shown in Figure 12.

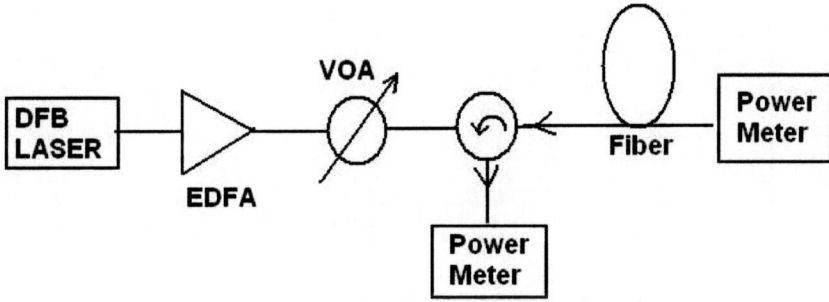

Figure 11. Experimental setup to measure threshold power in Stimulated Brillouin Scattering

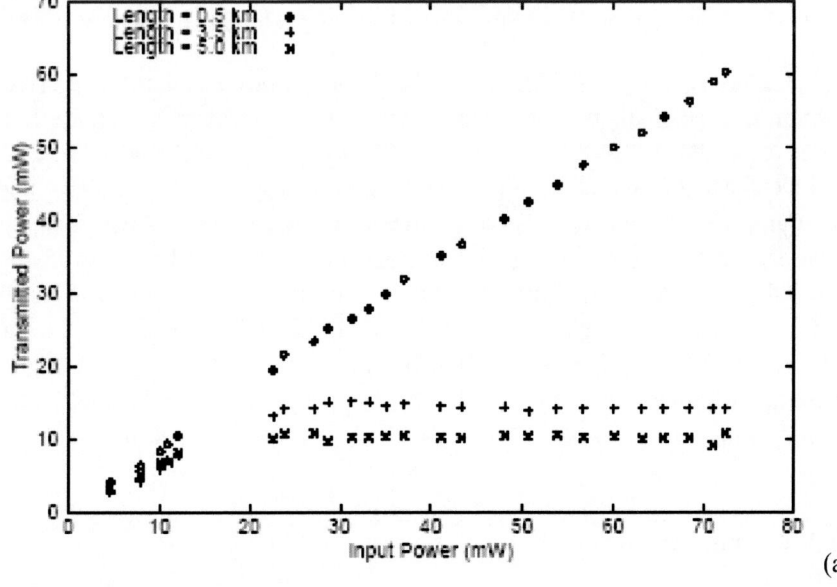

(a)

Figure 12. (Continued on next page)

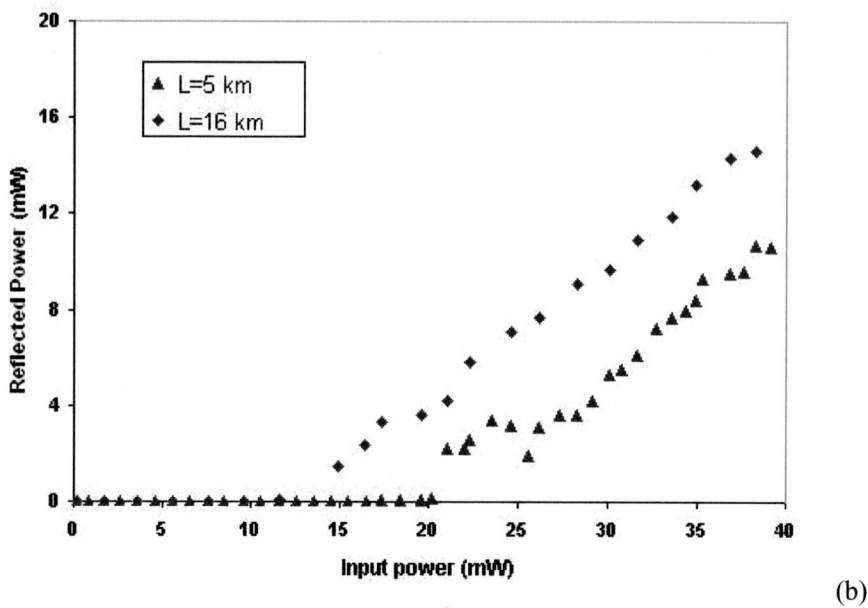

(b)

Figure 12. (a) Variation of transmitted power with input power for different lengths of DSF. (b) Variation of the reflected Stokes power with input for different lengths of DSF

For a shorter length of fiber, since the threshold power is large, the effective Brillouin gain is small and hence the conversion to Stokes power is not efficient. In this case, output power varies linearly with the input power. It can be concluded that, threshold for this fiber length occurs at much larger values of input pump power. This idea is corroborated by the fact that, as the fiber length increases, the Brillouin gain increases proportionately, thus bringing down the threshold power. Also, the throughput power measured for 5 km fiber is less than that for 3.5 km, indicating a larger conversion to the Stokes power. It is also seen from Figure 12(b) that the conversion efficiency of the Stokes power is larger for the longer length of the fiber. From these graphs, the SBS threshold can be determined, applying any of the conventional definitions stated earlier.

7.3.4. Numerical Model

In a single mode fiber of length L, the steady state equations describing the evolution of the pump and Stokes waves with position z along the fiber length are [3],

Specific Applications of Third Order Nonlinearity

$$\frac{dI}{dz} = -g_B BI - \alpha I \tag{59}$$

$$\frac{dB}{dz} = -g_B BI + \alpha B \tag{60}$$

where I and B are the pump and the Stokes intensities respectively, and α is the propagation loss in the fiber. These are two coupled differential equations and can be solved numerically using the standard Runge Kutta algorithm. It is more useful to write eqns (59) and (60) in terms of optical power and normalize them with respect to input power and fiber length [89]. If the input pump intensity is I_0, using $P(\zeta) = \frac{I}{I_0} = P$ and $S(\zeta) = \frac{B}{I_0} = S$, the coupled equations are

$$\frac{dP}{d\zeta} = -\sigma SP - \beta P \tag{61}$$

$$\frac{dS}{d\zeta} = -\sigma SP + \beta S \tag{62}$$

with $\zeta = z/L$, $\sigma = g_B L I_0$, $\beta = \alpha L$, while σ and β represent the normalized gain and loss factors respectively.

The above equations can be solved analytically for the case where the loss factor is zero. But for practical systems, since the loss factor cannot be neglected, these equations have to be solved numerically. The boundary conditions for solving the above normalized equations numerically are:

1. The pump input at $\zeta=0$ is 1
2. The Stokes power at $\zeta=1$ should be close to zero. The numerical value chosen for this noise input at the farthest end of the fiber decides the accuracy of the calculation.

The longitudinal step size is chosen so as to obtain converging solutions. The result of one such evaluation is shown in Figure 13.

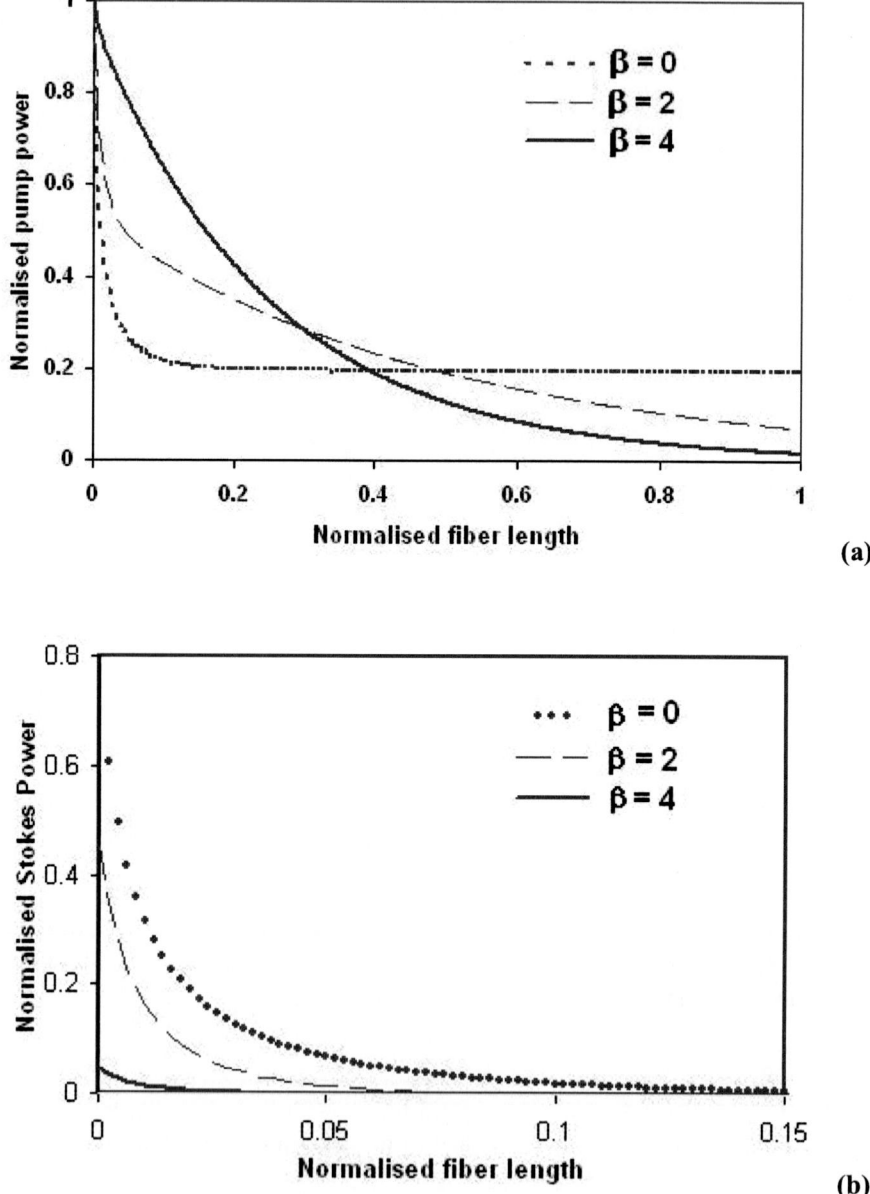

Figure 13. (a) Evolution of the normalized pump power with fiber length for different values of the loss factor (b) Evolution of the Stokes power with fiber length. Both the calculations were done for a gain factor of 100.

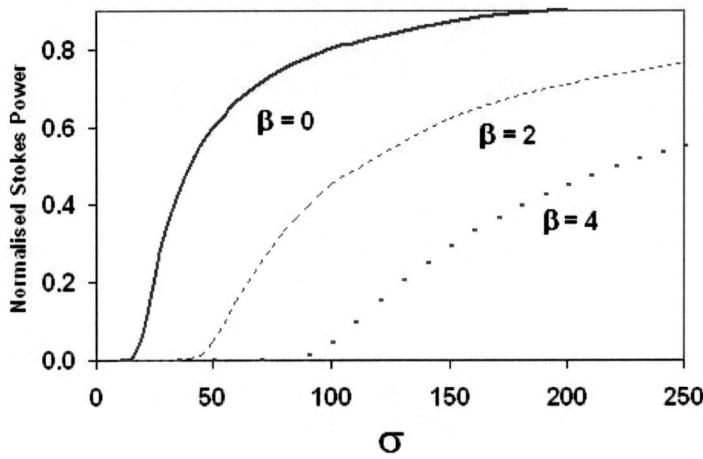

Figure 14. Variation of normalized Stokes power with gain factor, corresponding to different values of input pump power, computed for different values of loss factor.

Figure 13 shows the evolution of the pump and Stokes power along the length of the fiber for different values of loss factor. Considering propagation of pump power, P_0 through a standard single mode fiber, with effective area $50\,\mu m^2$, a gain factor σ of 100 would imply a $P_0 L$ product of 100. For instance, a span length of 50 km and an input power of the order of 2 mW would yield the value for σ as 100. Propagation with different values of loss factor is considered. $\beta = 0$ implies a propagation through an ideal lossless medium, and it is seen that the conversion efficiency is maximum in this case. For a large value of loss factor ($\beta = 4$), the pump power is below threshold and hence, the spatial evolution of the pump follows an exponential decay due to the fiber loss. The corresponding Stokes power is also found to be negligible. It is also evident from Figure 13 (b) that a significant growth of Stokes power occurs only at the input end of the fiber.

In order to simulate a practical situation, where the Brillouin threshold is calculated by increasing the input power and recording the backscattered Stokes power, the variation of Stokes power with σ needs to be considered. A typical result is shown in Figure 14.

It can be observed that the normalized Stokes intensity starts growing sharply beyond a particular value of σ, which corresponds to the threshold power. The rate of growth of Stokes power is highly dependent on the total loss in the fiber. By experimentally measuring the threshold power, and fitting to the above result, the Brillouin gain peak and hence, the acousto-optic characteristics of the fiber can be estimated.

7.3.5. Acousto-optic Effective Area

An optical fiber has a cylindrical symmetry and hence, there are three types of acoustic modes corresponding to longitudinal, torsional and flexural modes. Though each of these modes can cause acousto optic effects of their own, the lowest longitudinal acoustic mode interacts with the input pump photon and shows the strongest influence on Brillouin scattering [90,91]. The theory of SBS discussed above assumes that the optical mode interacts with the lowest order acoustic mode, and the area of interaction is approximated as the effective area of the optical mode. This approximation is not true in many practical situations. There could be large variations in the acoustic mode profile in different fibers, due to changes in refractive index profiles and dopant concentration.

Since SBS is the result of the interaction of optical wave with the material density fluctuations created by the wave, the equation for the acousto-optic interaction needs to be solved to obtain the exact form of density fluctuations, to obtain the various acoustic modes [92]. Each of the acoustic modes is assumed to interact independently with the optical field, and the propagation equation for the Stokes power is written in terms of an optical overlap factor. This overlap factor, also defined as acousto-optic effective area, is given by [93]

$$A_m^{ao} = \left[\frac{\langle f^2(r) \rangle}{\langle \xi_m(r) f^2(r) \rangle} \right]^2 \langle \xi_m^2(r) \rangle$$

where ξ_m is the mth order acoustic mode and $f(r)$ is the fundamental optical mode. The acousto-optic effective area is approximately equal to the optical effective area in the case of single mode fibers with step index profile. It has to be noted that, larger acousto optic effective area would imply that the acoustic modes are localized, resulting in a weaker overlap with the optical modes. This leads to smaller Brillouin gain coefficient and hence, an enhanced Brillouin threshold. Engineering a fiber design so as to increase the acousto-optic effective area is one of the recent trends to increase the Brillouin threshold in communication grade fibers [93].

Contrary to other nonlinear processes, the threshold pump power required for the onset of SBS is comparatively very low. SBS is known to severely limit the design of several optical systems, particularly broadband passive optical networks (B-PON) with analog video overlay since it places a relatively low limit on the optical power that can be launched in a single mode fiber. In addition to clamping the throughput power, SBS induced phase change degrades the optical signal to noise ratio. Analog systems are affected severely since these are loss-limited and require higher launch powers [94,95]. Since SBS gain depends on the line width of the pump, one of the sought-after techniques to counter SBS is to modulate the pump and increase its effective line width [88]. Other methods include the modification of the refractive index profile and the doping concentration, so as to increase the acousto optical effective area, and hence the threshold. At system level, the positioning of optical components are optimized for minimizing the noise due to SBS [96]. Use of a fiber Bragg grating whose stop band matches with the Stokes wavelength is another technique to avoid SBS [97]. Since strain and temperature can alter the acousto optic modes, both these parameters are also varied along the fiber to enhance the SBS threshold [98,99].

7.3.6. Applications of SBS

A wide variety of system and device applications are designed on the basis of SBS. Due to the narrow linewidth of Brillouin gain, and a relatively high value of gain coefficient, SBS is suitable for wavelength conversion, frequency selectivity, carrier suppression and amplification. One of the prominent applications of SBS is in its use in distributed temperature and strain sensors. Any environmental influence like temperature or strain alters the density fluctuations in the material, and hence the measurement of the Stokes power provides information on the temperature and mechanical stress on the fiber. Both the linewidth and the frequency of the Stokes wave are expected to change with temperature and strain [6].

Since the power from the pump wave is transferred to the Stokes wave, SBS can be used for amplifying signals which are counter-propagating with the pump at the Stokes frequency. The advantage of such a Brillouin amplifier is the lower threshold. However, this application is limited to narrow band digital signals and in radio-over-fiber applications due to the small bandwidth of Brillouin gain. In microwave photonic systems, the narrow linewidth of the Brillouin gain spectrum is utilized to selectively amplify a weak data carrying optical sideband, without perturbing an unmodulated carrier [100].

Design of a multiwavelength fiber laser has emerged as one of the attractive applications of SBS. In this design, a low power Stokes signal is given as the seed, and the backscattered Stokes signal from a long length of single mode fiber in the erbium doped fiber ring laser cavity is allowed to propagate within the cavity by an appropriate design. SBS has a cumulative action in this case, resulting in the generation of multiple wavelengths at the output of the laser cavity, which has significant applications in DWDM communication. The hybrid Brillouin Erbium Fiber Laser was first proposed by Cowle and Stepanov [101]. Since then, this configuration is used for multiple applications. In a self-seeded scheme, no external Brillouin pump is used and about 120 Stokes lines are generated with uniform powers [102]. The multiple wavelengths generated can be made tunable by including a Sagnac loop within the cavity, and by using polarization effects [103-106].

7.4. STIMULATED RAMAN SCATTERING

Raman scattering occurs when the incident light interacts with the vibrational modes of the molecules in the medium [3]. The resulting scattered light is downshifted in frequency, by an amount decided by the vibrational modes of the medium and is called as the Stokes wave. Quantum mechanically, Raman Effect is described as the inelastic scattering of a photon which creates or annihilates an optical phonon. It was first reported by C.V. Raman and K.S. Krishnan [107,108], and independently by Grigory Landsberg and Leonid Mandelstam in 1928. Though the process is similar to that of fluorescence, the underlying phenomena are entirely different. While fluorescence is a resonant effect, Raman scattering can occur at any frequency of incident light.

7.4.1. Theory

In the classical description, it is assumed that the optical polarizability of a molecule in a medium is not a constant, and is a function of the inter-nuclear distance. So, when the molecule is set into oscillation, its polarizability will be modulated in time, and hence, the refractive index of a collection of coherently oscillating molecules will be modulated in time. This temporal variation in refractive index would modify the incident light passing through the medium, resulting in a change in its frequency. This is *spontaneous* Raman scattering. The transformation efficiency of the process is usually low. In the presence of an optical field of frequency ω_p, the molecular vibrations modulate the refractive index of the medium at its resonant frequency of vibration, ω_v, thereby generating sidebands at frequencies $\omega_p \pm \omega_v$. Out of the generated frequencies, the Stokes wave at frequency $\omega_s (= \omega_p - \omega_v)$ can further beat with the laser field to produce a modulation of the total intensity, which coherently excites the molecular oscillations at the frequency, $\omega_v = \omega_p - \omega_s$. This interaction results in a stronger molecular vibration, which further leads to a generation of a stronger Stokes field. This cumulative action sets, in the presence of an intense incident field and the process is called *Stimulated* Raman Scattering (SRS). The corresponding antiStokes process is negligibly weak.

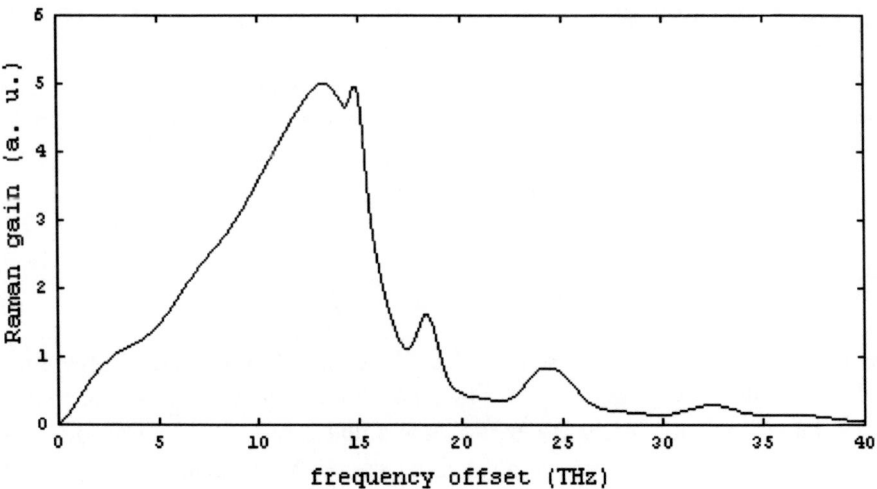

Figure 15. Raman gain against the frequency shift between the pump and Stokes wave in a fiber consisting of silica glass, showing a bandwidth of upto 40 THz and a broad peak near 13 THz [111].

In crystalline media, the resonant vibrational frequencies have a very narrow bandwidth and consequently, the generated Stokes and antiStokes wavelengths reflect the characteristics of the material. In an optical fiber, since silica is in the amorphous form, the resonant frequencies of the vibrational modes overlap and create a continuum, resulting in a Raman gain spectrum that is considerably broad

[109,110]. Typical Raman gain as a function of the frequency difference between the pump and the Stokes waves for silica glass in an optical fiber is shown in Figure 15.

The peak value of Raman gain is found to be about $3.1 \times 10^{-14} mW^{-1}$ for a pump wavelength of 1.55 μm [112]. Its value is found to decrease with the decrease in the pump wavelength. In a standard single mode fiber, the broad gain spectrum means that, if the launched pump power at a wavelength of 1.55 μm exceeds the Raman threshold, then the Stokes power is downshifted in frequency by about 13.2 THz, corresponding to a wavelength shift of about 112 nm. [112]. The peak gain is also a function of the dopant concentration. It is also interesting to note that Raman scattering does not have any phase matching condition as a pre-requisite. This is true for Brillouin scattering as well. But in dispersive fibers, there could be a walk-off between the pump and the Stokes waves in the case of short pulses, and hence, the Raman conversion efficiency depends on the temporal pulse duration.

Similar to the case of Brillouin scattering, Raman threshold is defined as the input pump power at which the output powers for pump and Stokes waves are equal. Assuming a Lorentzian shape for the Raman gain spectrum, it can be approximated as [6],

$$P_{th} \approx 16 \frac{K_S A_{eff}}{g_R L_{eff}} \tag{64}$$

where K_s is the factor which includes the relative polarization between the pump and the Stokes wave and g_R is the peak Raman gain. For a polarization maintaining fiber, $K_s = 1$, while it takes the value 2 for a standard single mode fiber. The threshold estimated from the above equation for a standard single mode fiber is about 2 W for an effective length of 25 km. Due to this large value of threshold, SRS is not expected to occur in single channel optical communication systems. However for WDM systems, SRS is found to play an active role, since the Stokes wave is already present in the system, and hence, the channels at longer wavelengths are amplified at the expense of channels at lower wavelengths. This also leads to degradation in signal-to-noise ratio and increase in bit error rates. Though the conversion efficiency within a band is found to be small, interaction between the C and L or S band is more efficient due to the gain peaking corresponding to those wavelength differences. In the case of pulse propagation, the effects due to GVD, SPM, XPM, pulse walk off and pump depletion need to be taken care of. Hence, the Nonlinear Schrödinger equations need to be written for the pump and the Stokes waves and these coupled amplitude equations need to be solved numerically [5].

SRS is found to be highly detrimental in high power lasers and amplifiers. A unique technique is demonstrated recently to counter the same with specially designed, ytterbium-doped fiber, with a germanium doped high index ring surrounding the core. This design can suppress SRS by coupling the Raman wavelength out of the fiber core through this ring [113].

7.4.2. Applications of SRS

Advances in Raman pump technology and demonstration of high power pumps required for Raman amplification have resulted in the utilization of Raman amplification at the system level. SRS can be used for signal amplification, if the wavelength difference between the pump and the signal is within the bandwidth of the Raman gain spectrum. Such amplifiers are called as Raman fiber amplifiers. The fact that the amplification depends on the separation between the pump and signal wavelengths, and not on the absolute value, is the distinguishing difference between Raman fiber amplifiers and other conventional amplifiers like EDFA and semiconductor optical amplifiers. The fundamental advantages of these amplifiers include amplification at any wavelength (assuming that the pump powers are available at corresponding wavelengths) and the absence of a specialized gain medium since the fiber itself is the active medium. In addition, the gain spectrum is decided by the pump spectrum and hence, there is a large amount of flexibility in designing such amplifiers. Broad band amplification is also possible with the use of larger number of pump wavelengths. SRS does not depend on the relative direction of propagation between the pump the signal and hence, forward, backward or bidirectional pumping schemes are widely used [5,6]

The fundamental principle of Raman amplification is used in distributed Raman amplifiers, discrete amplifiers and Raman fiber lasers. If the transmission fiber itself is used for amplification, it results in distributed Raman amplifiers (DRA). The DRA can make the system effectively repeater-less, and it reduces the power excursion in the signal, as against lumped amplifiers using EDFAs. Other deleterious nonlinear effects and the Amplified spontaneous emission (ASE) noise also can be minimized in DRA. By using multiple pumps at different wavelengths and appropriate power levels, a flat gain can be achieved in the different communication bands [114-116].

In order to amplify a signal at 1550 nm in the telecom window, a pump signal at a wavelength close to 1450 nm is required. Lack of high power lasers and poor efficiency are some of the major challenges in Raman amplification [117]. Other issues include safety while handling high power and optical damage, which include fiber fuse, burning due to tight bending and connector damage. Since the Raman scattering cross section and consequently the Raman gain is poor in standard single mode fibers, Raman fibers are designed with germano-silicate and phosphor-silicate glasses, where an enhanced Raman gain is observed [118]. The sources of noise in a Raman amplifier include the amplified spontaneous emission due to spontaneous Raman Scattering, signal – spontaneous beat noise, multiple path interference due to discrete sources of reflection and due to Rayleigh backscattering, pump-pump four wave mixing, pump-signal four wave mixing, polarization dependent gain, pump mediated signal cross talk, and noise transfer from pump to signal [119].

One of the interesting applications of lumped Raman amplification is in coarse wavelength division multiplexing transmission systems which require a large bandwidth of about 100 nm. Enhanced Raman gain is demonstrated in such systems with the use of highly nonlinear fiber (HNLF) [120]. Dispersion compensating fibers and dispersion compensating photonic crystal fibers are also expected to have enhanced Raman gain [121,122]. Holey fibers, consisting of pure silica surrounded by a regular array of longitudinal air holes, have also been demonstrated to provide enhanced Raman efficiency [123,124]. The optical signal-to-noise ratio is also expected to improve when a piece of highly nonlinear photonic crystal fiber (PCF) is inserted at the beginning of the link, in a backward pumped Raman amplifer [125].

A Raman fiber laser is basically a Raman amplifier in a resonator cavity. A fiber Bragg grating is usually used as the resonating device. Since the field propagating in the cavity is large, the effective threshold is decreased. Due to the large spectral width of the Raman amplifier, the Raman laser can be tuned to a wide range of wavelengths. Cascaded Raman effect can be used where the generated Stokes wavelength itself acts as the pump and thus, any desired wavelength can be generated. The cascaded effects are also used to construct grating-free Raman lasers, by using a broad band pump from an Yb^{3+} doped fiber laser [126]. Raman fiber laser finds application in sensors which use Fiber Bragg grating as the sensing element, and the wavelength shift is amplified using the wide bandwidth Raman fiber lasers [127].

7.5. SOLITON LASERS

Nonlinear and dispersive effects are detrimental in conventional optical fiber communication. However, a new class of optical transmission system called the optical Soliton Fiber communication system has been demonstrated, which utilizes the dispersion and nonlinear effects for the creation and propagation of robust solitons at ultra-high speeds and without electronic repeaters [128]. Optical soliton fiber laser is a nonlinear optical oscillator, containing optical fibers, which generates optical pulses due to a balance between various sources of linear and nonlinear gain, loss, self phase modulation and dispersion [129]. Soliton formation plays a major role in the shortening of the pulses in a soliton laser. For high-bit-rate TDM systems, the pulse duration should be very short. For instance, the pulse duration of a 100 Gbit/s system should be shorter than a few picoseconds. These short duration pulses are rich in spectral content and hence can also be used as sources in multichannel WDM systems with the use of appropriate filters. Ultra short laser pulse technology also finds applications in diverse fields like laser processing, medical and bio optics, opto-electronics and spectroscopy.

The simplest scheme to make a soliton laser is to attach an auxiliary cavity having a fiber with anomalous dispersion to a mode-locked laser. Multiple passes of the pulse in the auxiliary cavity results in pulse shortening. The most commonly used cavity geometries are linear, ring, figure-of-eight and sigma cavities [130-134]. A nonlinear polarization rotation (NPR) laser is made by having an erbium doped fiber amplifier in a ring cavity structure, with polarisers and polarization controllers. The dispersion in the fiber cavity is deliberately chosen to be low, so that, after one round-trip propagation, different parts of the pulse experience different values of nonlinear phase, leading to a large rotation in the polarization direction. The polarization controller and the polarizer allow the high intensity peak to pass through, thus effectively shortening the pulse in each round trip [135].

In a stretched pulse NPR laser, the dispersion is managed in the fiber cavity by introducing both normal and anomalous dispersion fibers in the cavity, so that the average dispersion is zero. The high anomalous dispersion fiber is placed before the EDFA, so that the pulse broadens and hence gets amplified better. The amplified pulse is further allowed to pass through a normal dispersion fiber, which compresses the pulse. The nonlinearity-induced phase change is much less in this case due to pulse stretching while the overall pulse energy available at the output can be enhanced [136-137]. The longer cavity lengths make the pulse repetition rate slower.

Another configuration to achieve soliton laser is the figure-of-eight fiber laser. It consists of a twin cavity; one containing the erbium doped fiber amplifier with isolator. A part of the power from this cavity is coupled to a nonlinear optical loop mirror consisting of a dispersion shifted fiber, which acts as a saturable absorber allowing high powers, while blocking low power components of the pulse. Figure-of-eight laser is also demonstrated with a dispersion compensating fiber (DCF) in the loop along with the DSF, thus making it a dispersion imbalanced loop mirror. This configuration helps to shorten the pulse without any CW background [138]. Active mode-locking with the help of an optical modulator in the cavity can also be used for the generation of short pulses. They are found to be environmentally more stable than the passive mode locking schemes. In the presence of nonlinearity and dispersion, spectral broadening and temporal shortening can occur in these types of lasers [131,139].

A sliding frequency soliton laser has an acousto-optic modulator in the cavity, which shifts the frequency of the propagating wave in each round trip. This laser has a relatively low self starting threshold as CW lasing and noise are suppressed by the frequency shifter and a filter included in the cavity [140,141]. There is an asymmetrical spectral loss in each round trip which is balanced through self phase modulation. Another intelligent design to achieve large pulse energy is to have nonlinearity management instead of dispersion management by using self defocusing effect in certain materials [142]. The soliton output from any of these designs are shown to be either normal solitons, bound solitons or in a noise-like pulse state.

Tunable lasers are required for spectroscopic applications. Ultra wide tunable soliton laser is generated and these high energy solitonic pulses are further used to demonstrate ultra flat supercontinuum, with low noise figure which is highly coherent and suitable for frequency comb generation, using highly nonlinear photonic crystal fibers with small dispersion [143]. The subject of soliton lasers and soliton transmission systems is an area of active research.

7.6. SUPERCONTINUUM GENERATION IN PHOTONIC CRYSTAL FIBERS

Conventional telecom fibers are made of a doped solid silica core with a refractive index more than that of the undoped silica cladding. On the contrary, Photonic crystal fibers (PCFs) are made from a single material with a spatially periodic pattern of air holes surrounding the core region. These air holes will be present all along the length of the fiber. In an index-guiding PCF, the core is of solid silica. The air-silica cladding has an effective index lower than that of the silica core and hence the guidance of light is similar to the conventional fibers. In photonic band gap (PBG) PCFs, the core is an air-hole surrounded by an air-silica cladding. The index of the cladding is thus more than that of the core and the guidance of light is by the PBG effect. When the wavelength dependence of the cladding index counters the wavelength dependence of the V-parameter of the fiber with an appropriate arrangement of air-holes in the cladding, the fiber remains single-moded over the visible and near-IR range of wavelengths [144].

Supercontinuum (SC) generation, in the context of optical fibers, refers to the generation of a broadband light due to the propagation of short-duration pulses of high intensity through the fiber medium. When the wavelength of the laser pulses is close to the zero-dispersion wavelength of the fiber medium, the extent of SC generated is very large. Even though broadband generation is possible with standard fiber designs such as a dispersion-decreasing fiber or a dispersion flattened fiber [145], as well as by tapering the standard fibers [146], SC generation spanning more than 1000 nm was demonstrated when the zero-dispersion wavelength of the PCF was close to the wavelength of Ti:Sapphire femtosecond laser [147]. The impact of the nonlinearity of the fiber, manifested dominantly through self-phase modulation and four wave mixing, is significant when the dispersion effects are minimized, in combination with the presence of high peak powers [148,149]. The co-operative effect of a large number of nonlinear effects results in the generation of a wide set of new frequencies. The continuum is the result of all these new frequencies which are equi-spaced in the spectral domain as decided by the repetition rate of the incident laser pulses.

The input pulse duration, intensity and the repetition rate, along with the material characteristics of the fiber such as the zero dispersion wavelength and the nonlinear coefficient primarily decide the efficiency of the process of SC generation. As mentioned earlier, since the PCF is single-moded at a wide range of wavelengths, spatial coherence of the SC at the output is ensured. The enhanced spectral broadening can result in reduced temporal coherence. A source based on SC generation finds applications in areas ranging from WDM communication systems [150], multi-wavelength characterization of fiber-optic components, optical metrology [151], to optical coherence tomography in medicine [152].

Chapter 8

SUMMARY

This book gave an overview of some of the methods by which the third order susceptibility of optical fibers can be used for designing specific applications. The trigger for the study of various nonlinear phenomena in optical fibers is the availability of high power sources and optical amplifiers. Even though a large section of the existing literature is covered, the list is not exhaustive. Those nonlinear phenomena prominent in optical fibers, and which are of direct consequence to optical communication systems, are dealt with. Different aspects of these phenomena are examined, with the thrust on their applications. The list of applications in many cases is not comprehensive. Some applications like nonlinear optical phase conjugation, which can be used for the compensation of optical distortion, are not addressed. There are also some non-conventional applications of optical nonlinearity in the design of all-optical gates and optical computing, which are not discussed. Use of high powers is indispensable for long haul communication systems, and nonlinear effects are inevitable in these designs. An intelligent system design, which makes use of the nonlinear effects favorably, would cater to the demands of futuristic communication systems.

ACKNOWLEDGEMENTS

The authors gratefully acknowledge the assistance of Dilip D. Shah and J. Ravikanth in the experimental and modeling work. A large part of the authors' work reviewed here was sponsored by Department of Information Technology, Government of India.

REFERENCES

[1] Y. R. Shen, *Principles of Nonlinear Optics;* John Wiley & Sons: Singapore, 1991.
[2] R. Menzel, *Photonics: Linear and non-linear interactions of laser light and matter;* Springer-Verlag: Berlin 2001.
[3] R. W. Boyd, *Nonlinear Optics;* Academic Press: San Deigo, CA, 2003; 2nd Edn.
[4] N. Bloembergen, *Nonlinear Optics;* World Scientific: Singapore, 1996 4th Edn.
[5] G. P. Agrawal, *Nonlinear Fiber Optics;* Academic Press: San Diego, CA, 2001; 3rd Edn.
[6] T. Schneider, *Nonlinear Optics in Telecommunications;* Springer-Verlag: Berlin, 2004.
[7] Kleinman, D.A; *Phys. Rev.* 1962,126, 1977-1979.
[8] Broderick, N.G.R; Monro, T.M; Bennett, P.J; Richardson, D.J; *Opt. Lett,* 1999, 24, 1395-1397.
[9] Hiroishi, J; Sugizaki, R; Aso, O; Tadakuma, M; Shibuta, T; *Furukawa Review* 2003, 23, 21-25.
[10] Harbold, J.M; Ilday, F.O; Wise, F.W; Sanghera, J.S; Nguyen, V.Q; Shaw, L.B; Aggarwal, I.D; *Opt. Lett.* 27, 119-121 (2002).
[11] Gordon, J.P; *Opt. Lett.* 1986, 11, 662-664.
[12] Atieh, K; Myslinski, P; Chrostowski, J; Galko, P; *J. Lightwave. Technol.* 1999,17, 216-221.
[13] Sinkin, O.V; Holzlohner, R; Zweck, J; Menyuk, C.R, *J. Lightwave. Technol.* 2003, 21, 61-68.
[14] Bosco, G; Carena, A; Curri, V; Gaudino, R; Poggiolini, P; Benedetto, S; *IEEE Photon. Technol. Lett.* 2000, 12, 489-491.

[15] Atieh, K; Myslinski, P; Chrostowski, J; Galko, P; *Opt. Commun.* 1997,133, 541-548.
[16] Chernikov, S.V; Taylor, J.R; Mamyshev, P.V; Dianov, E.M; *Electron. Lett.* 1992, 28, 931-932.
[17] G.P.Agrawal, *Applications of Nonlinear Fiber Optics*, Academic Press: Boston, 2001.
[18] G. P. Agrawal, *Fiber-optic communication systems*, John Wiley & Sons, Inc., NewYork, 3rd Edn. (2002).
[19] Special issue on Multiwavelength fiber optic communication. *IEEE Communication magazine,* 1998, 36, 26-68.
[20] Nakao, M; Nishida S.K; Tamamura, T; *IEEE J. Sel. Areas Commun.,* 1990, 8, 1178-1182.
[21] Zah, C.E; Lin, P.S.D; Favire, F; Pathak, B; Bhat, R; Caneau, C; Gozdz, A.S; Andreadakis, N.C; Koza, M.A; Lee, T.P; Wu, T.C; Lau, K.Y; *Electron. Lett.* 1992, 28, 824-825.
[22] Zah, C.E.; Amersfoort, M.R.; Pathak, B.; Favire, F.; Lin, P.S.D.; Rajhel, A.; Andreadakis, N.C.; Bhat, R.; Caneau, C.; Koza, M.A. *IEEE Photon. Technol. Lett.* 1996, 8, 864-866.
[23] Schlager, J. B; Kawanishi, S; Saruwatari. M; *Electron. Lett.* 1991, 27, 2072-2073.
[24] Takara, H.; Kawanishi, S.; Saruwatari, M; Schlager, *J. B. Electron. Lett.* 1992, 28, 2274-2275.
[25] Monnard, R.; Doerr, C.R.; Joyner, C.H; Zirngibl, M.; Stulz, L.W. *IEEE Photon. Technol. Lett.* 1997, 9, 815-817.
[26] Doerr C.R.; Joyner, C.H.; Stulz, L.W.; Centanni, J.C. *IEEE Photon. Technol. Lett.* 1997 9, 1430-1432.
[27] Zirngibl, M. *IEEE Commun. Mag.*1998, 36, 39-41.
[28] Reeve, M. H.; Hunwicks, A. R.; Zhao, W.; Methley, S. G.; Bickers, L.; Hornung, S. *Electron. Lett.* 1988, 24, 389-390.
[29] Wagner, S. S.; Chauran, T. E. *Electron. Lett.* 1990, 26, 696-697.
[30] Lee, J. S.; Chung, Y. C.; DiGiovanni, D. J. *IEEE Photon. Technol. Lett.* 1993, 5, 1458-1461.
[31] Chung, Y. C.; Lee, J. S.; Derosier, R. M.; DiGiovanni, D. J. *Electron. Lett.* 1994,30, 1427-1428.
[32] Morioka, T.; Mori, K.; Saruwatari, M. *Electron. Lett.,* 1993, 29, 862-864.
[33] Morioka, T.; Mori, K.; Kawanishi, S.; Saruwatari, M. *IEEE Photon. Technol. Lett.* 1994, 6, 365-367.
[34] Morioka, T.; Kawanishi, S.; Mori, K.; Saruwatari, M. *Electron. Lett.* 1994, 30, 790-791.

[35] Morioka, T.; Kawanishi, S.; Mori, K.; Saruwatari, M. *Electron. Lett.* 1994, 30, 1166-1168.

[36] Morioka, T.; Uchiyama, K.; Kawanishi, S.; Suzuki, S.; Saruwatari, M. *Electron. Lett.* 1995, 31, 1064-1066.

[37] Takushima, Y.; Futami, F.; Kikuchi, K. *IEEE Photon. Tech. Lett.* 1998, 10, 1560-1562.

[38] Takushima, Y.; Kikuchi, K. *IEEE Photon. Tech. Lett.* 1999, 11, 322-324.

[39] Mikulla, B.; Leng, L.; Sears, S.; Collings, B. C.; Arend, M.; Bergman, K. *IEEE Photon. Technol. Lett.* 1999, 11, 418-420.

[40] Boyraz, O.; Kim, J.; Islam, M. N.; Coppinger, S.; Jalali, B. *J. Lightwave Technol.* 2000, 18, 2167-2175.

[41] Boivin, L.; Taccheo, S.; Doerr, C. R.; Stulz, L. W.; Monnard, R.; Lin, W.; Fang, W. C. *IEEE Photon. Technol. Lett.* 2000, 12, 1695-1697.

[42] De Souza, E. A.; Nuss, M. C.; Knox, W. H.; Miller, D. A. B. *Opt. Lett.* 1995, 20, 1166-1168.

[43] Nuss, M. C.; Knox, W. H.; Koren, U. *Electron. Lett.* 1996, 32, 1311-1312.

[44] Tamura, K.; Yoshida, E.; Nakazawa, M.; *Electron. Lett.* 1996, 32, 1691-1693.

[45] Tamura, K.; Yoshida, E.; Yamada, E.; Nakazawa, M.; *Electron. Lett.* 1996, 32, 835-836.

[46] Guy, M. J.; Chernikov, S. V.; Taylor, J. R. IEEE Photon. *Technol. Lett.* 1997, 9, 1017-1019.

[47] Takushima, Y.; Futami, F.; Kikuchi, K. IEEE Photon. *Technol. Lett.* 1998, 10, 1560–1562.

[48] Nakazawa, M.; Tamura, K.; Kubota, H.; Yoshida, E. *Opt. Fiber Technol.* 1998, 4, 215–223.

[49] Takushima, Y.; Kikuchi, K. IEEE Photon. *Technol. Lett.* 1999, 11, 322–324.

[50] Tamura, K.R.; Kubota, H.; Nakazawa, M. *IEEE J. Quantum Electron.* 2000, 36, 773–779.

[51] Boivin, L.; Wegmeuller, M.; Nuss, M.C.; Knox, W. H. *IEEE. Photon. Technol. Lett.* 1999, 11, 466-468.

[52] Ravikanth, J.; Shah, D.D.; Vijaya, R.; Singh, B.P.; Shevgaonkar, R.K.; *Microwave Opt. Technol. Lett.* 2002, 32, 64–70; *ibid*, 2002, 34, 399 (erratum).

[53] Shah, D.D.; Dixit, Nimish; Vijaya, R.; *Opt. Fib. Technol.* 2003, 9, 149-158.

[54] Deepa, R.; Vijaya, R. (communicated).

[55] Hill, K.O.; Johnson, D.C.; Kawasaki, B. S.; Mac-Donald, R. I.; *J. Appl. Phys.* 1978,49, 5098-5106.
[56] Shibata, N.; Braun, R.P.; Waarts, R. G.; *IEEE J. Quant. Electron.* 1987, 3, 1205-1210.
[57] Inoue, K.; Toba, H. *IEEE Photon. Technol. Lett.* 1992,4, 69-72.
[58] Inoue, K.; *J. Lightwave Technol.* 1992,10, 1553-1561.
[59] Watanabe, S.; Chikama, T. *Electron. Lett.* 1994,30, 163-164.
[60] Inoue, K.; Toba, H. *J. Lightwave Technol.* 1995,13, 88-93.
[61] Tkach, R. W.; Chraplyvy, A.R.; Forghieri, F.; Gnauck, A. H.; Derosier, R. M. *J. Lightwave Technol.* 1995,13, 841-849.
[62] Gao, S.;Yang, C.; Jin, G. *Opt. Commun.* 2002,206, 439-443.
[63] Inoue, K.; Toba, H. IEEE Photon. *Technol. Lett.* 1992,4, 69-72.
[64] Sefler, G.A.; Kitayama, K; *J. Lightwave Technol.* 1998,16, 1596-1605.
[65] Yamamoto, T.; Nakazawa, M. IEEE Photon. *Technol. Lett.* 1997,9, 327-329.
[66] Song, S.; Allen, C.T.; Demarest, K. R.; Hui, R. *J. Lightwave Technol.*1999,17, 2285-2289.
[67] Deepa, R; Vijaya, R; *Opt. Commun.* 2007, 269, 206-214.
[68] Inoue, K.; *J. Lightwave Technol.*1993, 11, 2116-2122.
[69] Inoue, K; *IEEE Photon Technol Lett.*1992, 4, 888-890.
[70] Neokosmidis, T. Kamalakis, A. Chipouras, T. Sphicopoulos, *J. Lightwave. Technol.* 2005, 23,1137-1143.
[71] Inoue, K.; J. Lightwave. *Technol,* 1993,11, 455-461.
[72] Forghieri, F.; Tkach, R. W;. Charplyvy, A. R. *J. Lightwave. Technol.* 1995, 13, 889-897.
[73] Bogoni, A.; Poti, L.; *IEEE J. Sel. Topics in Quant. Electr.* 2004, 10, 387-392.
[74] Asu, O.; Tadakuma, M.; Namiki, S. Furukawa Review, 2000,19, 63-68.
[75] Curti, F.; Matera, F.; Maria T-Beleffi, G.; *Opt. Commun.,* 2002,208, 85-89.
[76] Blows, J. L. *Opt. Commun.* 2004, 236, 115-122.
[77] Sarker, B.C.; Yoshino, T.; Majumder, S. P.; *Optik.*2003, 113, 541-547.
[78] Islam, M. N.; Boyraz, O. *IEEE J. Sel. Topics in Quant. Electr.* 2002, 8, 527-537.
[79] Chen, H. *Opt. Commun.* 2003,220, 331-335.
[80] Eiselt, M.; Jopson, R. M.; Stolen, R. H. *J. Lightwave. Technol.* 1997,15, 135-143.
[81] Lin, Q.; Agrawal, G. P. IEEE Phot. *Technol. Lett.* 2003, 15, 1719-1721.
[82] Ciaramella, E.; Trillo, S. *IEEE Photon. Technol. Lett.* 2000,12, 849-851.

References

[83] Ciaramella, E.; Curti, F.; Trillo, S.; *IEEE Photon. Technol. Lett.* 2001, 13, 142-144.

[84] Yamamoto, T.; Nakazawa, M. *IEEE Photon. Technol. Lett.* 1997, 9, 1595-1597.

[85] Aoki, Y.; Tajima, K.; Mito, I. *IEEE J.Lightwave.Technol.* 1988,6, 710-719.

[86] Gaeta, A.; Boyd, R.W. *Phys. Rev.* A, 1991,44, 3205-3209.

[87] Bao, X.; Brown, A.; DeMerchant, M.; Smith, *J.Opt. Lett.* 1999,24, 510-512.

[88] Billington, R. *National Physical laboratory Report,* 1999, COEM 31, 1-34.

[89] Bayvel, P.;Radmore, P. M. *Electron. Lett.* 1990, 26, 434-436.

[90] Buckland, E. L., Boyd, R. W. *Opt. Lett.*1997, 22, 676-678.

[91] Thomas, P. J.; Rowell, N. L.; Van Driel, M. G.; Stegeman, I. *Phys. Rev. B.*1979, 19, 4986-4998.

[92] Koyamada, Y.; Sato, S.; Nakamura, S.; Sotobayashi, H.; Chujo, W. *J. Lightwave. Technol.* 2004, 22, 631-639.

[93] Kobyakov, A.; Kumar, S.; Chowdhury, D. Q.; Ruffin, A. B.; Sauer, M.; Bickham, S. R.; Mishra, R. *Opt. Exp.* 2005,13, 5338-5346.

[94] Peral, E.; Yariv, A. *IEEE J. Quant. Electron.* 1999, 35, 1185-1195.

[95] Horowitz, M.; Chraplyvy, A.R.; Tkach, R.W.; Zyskind, J.L. *IEEE Photon. Technol. Lett.*1997, 9, 124-126.

[96] Hu, L.; Kaszubowska, A.; Barry, L. *Opt. Commun.* 2005, 255,253-260.

[97] Lee, H.; Agrawal, G. P.; *Opt. Exp.* 2003,11, 3467-3472.

[98] Yoshizawa, N.; Imai, T.; *J. Lightwave. Technol.* 1993, 11, 1518-1522.

[99] Imai, Y.; Shimada, N. *IEEE Photon. Technol. Lett.* 1993, 5, 1335-1337.

[100] Yao, X.S. IEEE Photon. *Technol. Lett.,* 1998, 10, 138-140.

[101] Cowle, G. J., Stepanov, D. Yu. *IEEE Photon. Technol. Lett.* 1996, 8, 1465-1467.

[102] Song, Y. J.; Zhan, L.; Ji, J. H.; Su, Y.; Ye, Q. H.; Xia, Y. X. *Opt. Lett.* 2005, 30, 486-488.

[103] Song, Y. J.; Zhan, L.; Hu, S.; Ye, Q. H.; Xia, Y. X. *IEEE Photon. Technol. Lett.* 2004, 16, 2015-2017.

[104] Al Mansoori, M. H.; Kamil Abd-Rahman, M.; Adikan, F. R. M.; Mahdi, M.A. *Opt. Exp.* 2005, 13, 3471-3476.

[105] Fok, M. P.; Shu, C.; *Opt. Exp.* 2006, 14, 2618-2624.

[106] Abdullah, M. K., Shaharudin, S.; Mahdi, M. A.; Endut, R.; *Opt. Laser Technol.* 2004, 36, 567-570.

[107] Raman, C. V.; Krishnan, K.S. *Nature.* 1928, 121,501-502.

[108] Raman, C. V.; *Ind. Jour. Phy.*1928, 2, 388-398.
[109] Stolen, R. H.; Gordon, J. P.; Tomlinson, W. J.; Haus, H. A. *J. Opt. Soc. Am. B,* 1989,6, 1159-1166.
[110] Shuker, R.; Gammon, R. W. *Phys. Rev. Lett.* 1970,25, 222-225.
[111] Hollenbeck, D.; Cantrell, C. D. *J. Opt. Soc. Am B* 2002,19,2886-2892.
[112] Mahgereftesh, D; Butler, D.L; Goldhar, J; Rosenberg, B; Burdge, G.L; *Opt. Lett.,* 1996, 21, 2026-2028.
[113] Fini, J.M; Mermelstein, M.D; Yan, M.F; Bise, R.T; Yablon, A.D; Wisk, P.W; Andredjco, M.J; *Opt. Lett.* 2006, 31, 2550-2552.
[114] Hu, J.; Marks, B. S.; Menyuk, C. R.; *J. Lightwave. Technol.* 2004, 22, 1519-1522.
[115] Perlin, V.; Winful, H.; *J. Lightwave. Technol.* 2002, 20, 250-254.
[116] Kidorf, H.; Rottwitt, K.; Nissov, M.; Ma, M.; Rabarijana, E. *IEEE Photon. Technol. Lett.* 1999, 11, 530-532.
[117] Namiki, S.; Seo, K; Tsukiji, N.; Shikii, S. *Proc. IEEE,* 2006, 94, 1024-1034.
[118] Dianov, E. M.; *J. Lightwave. Technol.* 2002, 20, 1457-1462.
[119] Bromage, J.; *J. Lightwave. Technol.* 2004, 22, 79-93.
[120] Miyamoto, T.; Tanaka, M.; Kobayashi, J.; Tsuzaki, T.; Hirano M., Okuno, T.; Kakui, M.; Shigematsu, M.; *J. Lightwave. Technol.* 2005, 23, 3475-3483.
[121] Bromage, J.; Rottwitt, K.; Lines, M. E. IEEE Photon. *Technol. Lett.* 2002, 14, 24-26.
[122] Varshney, S. K.; Saitoh, K.; Koshiba, M. IEEE Photon. *Technol. Lett.* 2005, 17, 2062-2064.
[123] De Matos, C. J. S.; Hansen, K. P.; Taylor, J. R. *Electron Lett.* 2003,39, 424-425.
[124] Yussof, Z.; Lee, J. H.; Belardi, W.; Monro, T. M.; The, P.C.; Richardson, D. J. *Opt. Lett.* 2002, 27, 424-426.
[125] Zhao, C-L.; Li Z.; Yang, X.; Lu, C.; Jin, W.; Demokan, M.S. *IEEE Photon. Technol. Lett.* 2005, 17, 561-563.
[126] Zhao, Y.; Jackson, S. *Opt. Exp* 2005, 13, 4731-4736.
[127] Peng P-C.; Tseng, H-Y.; Chi, S. *IEEE Photon. Technol. Lett.* 2004, 16, 575-577.
[128] Hasegawa, A.; *Opt. Lett.*1983, 8 650 (1983).
[129] K. Porsezian, V. C. Kuriakose, *Optical Solitons: Theoretical and Experimental Challenges,* Springer-Verlag, 2003.
[130] Smith, K.; Armitage, J. R.; Wyatt, R.; Doran, N. J.; Electron. Lett. 1990, 26, 1149-1151.

[131] Kafka, J.D; Bear, T; Hall, D.W; *Opt. Lett.*1989, 14, 1269-1271.
[132] Carruthers, T.F; Duling III, I.N; *Opt. Lett,* 1996, 21, 1927- 1929.
[133] Duling III, I.N; *Electron Lett.* 1991, 27, 544- 545.
[134] Richardson D. J.; Laming, R. I; Payne, D. N.; Philips, M.W.; Matsas, V.J. *Electron. Lett.* 1991, 27, 730-732.
[135] Hasegawa A., Matsumoto M., *Optical Solitons in Fibers,* Springer Series in Photonics, Berlin, 2003 3rd Edn.
[136] Haus, H.A.; Tamura, K.; Nelson, L.E.; Ippen, E.P. *IEEE J. Quantum Electron.* 1995, 31, 591-598.
[137] Tamura, K.; Ippen, E. P.; Haus, H. A.; Nelson, L. E; *Opt. Lett.* 1993, 18, 1080-1082.
[138] Gong, Y. D.; Shum, P.; Tang, D. Y.; Lu, C.; Gao, X.; Paulose, V.; Man, W. S.; Tam, H. Y. *Optics and Laser Technol.* 2004, 36, 299-307.
[139] Kartner, F. X.; Kopf, D.; Keller, U.; *J. Opt. Soc. Am.* B,1985. 12, 486-496.
[140] Sabert, H.; Brinkmeyer, E. *Electron.Lett.* 1993, 29, 2122-2124.
[141] Fontana, F.; Bossalini L; Franco, P.; Midrio, M.; Romagnoli, M.; Wabnitz, S.; *Electron. Lett.* 1994,30,321-.322.
[142] Ilday, F. O., Wise, F. W., *J.Opt. Soc.Am.* B, 2002, 19, 470-476.
[143] Takayanagi, J.; Nishizawa, N. *Jpn. J. Appl. Phys.* 2006,45, L441-L443.
[144] Birks, T.A; Knight, J.C; Russell, P.St.J; *Opt.Lett.* 1997, 22, 961-963.
[145] Okuno, T; Onishi, M; Nishimura, M; *IEEE Photon.Tech.Lett.* 1998, 10, 72-74.
[146] Birks, T.A; Wadsworth, W.J; Russell, P.St.J; *Opt.Lett.* 2000, 25, 1415-1417.
[147] Genty, G; Lehtonen, M; Ludvigsen, H; Kaivola, M; *Opt.Exp.* 2004, 12, 3471-3480.
[148] Tse, M.L.V; Horak, P; Poletti, F; Broderick, N.G.R; Price, J.H.V; Hayes, J.R; Richardson, D.J; *Opt.Exp.* 2006, 14, 4445-4451.
[149] Coen, S; Chau, A.H.L; Leonhardt, R; Harvey, J.D; Knight, J.C; Wadsworth, W.J; Russell, P.St.J;. *J.Opt.Soc.Am.* B 2002, 19, 753-764.
[150] Morioka, T; Takara, H; Kawanishi, S; Kamatani, O; Takiguchi, K; Uchiyama, K; Saruwatari, M; Takahashi, H; Yamada, M; Kanamori, T; Ono, H; *Electron. Lett.* 1996, 32, 906-907.
[151] Cundiff, S.T; Ye, J; Hall, J.L; *Rev.Sci Instrum.* 2001, 72, 3749-3771.
[152] Hartl, I; Li, X.D; Chudoba, C; Ghanta, R.K; Ko, T.H; Fujimoto, J.G; Ranka, J.K; Windeler, R.S; *Opt. Lett.* 2001, 26, 608-610.

INDEX

A

Aβ, 13, 19
Abdullah, 75
absorption, 15, 21, 29
accuracy, 18, 19, 20, 21, 53
acoustic, 47, 48, 49, 56
acoustic waves, 47, 48
air, 62, 64
algorithm, 18, 53
allocated time, 33
alternatives, 34
alters, 2, 57
amorphous, 6, 59
amplitude, 10, 11, 13, 15, 27, 38, 61
analog, 57
annihilation, 48
anomalous, 16, 24, 26, 31, 32, 40, 45, 63
application, 6, 47, 57, 62
argument, 28
assumptions, 14
asynchronous, 35
atoms, 5
attention, 34
availability, 67

B

backscattered, 50, 55, 58
backscattering, 62
band gap, 65
bandwidth, 2, 3, 14, 33, 46, 57, 59, 61, 62
beams, 1
beating, 35
behavior, 15, 30
bending, 62
birefringence, 15
Boston, 72
boundary conditions, 53
Bragg grating, 62
broad spectrum, 14
broadband, 34, 35, 57, 65
burning, 62

C

capacity, 2, 34
carrier, 15, 27, 33, 46, 57
cavities, 63
channels, 3, 12, 20, 33, 34, 35, 46, 60
cladding, 64
classical, 58
clouds, 5
coherence, 65
commercial, 3
communication, vii, 2, 3, 4, 6, 10, 12, 34, 35, 46, 50, 56, 58, 60, 61, 63, 65, 67, 72
communication systems, vii, 2, 3, 4, 12, 34, 35, 50, 60, 65, 67, 72

Index

compensation, 67
components, 1, 6, 11, 14, 15, 20, 26, 27, 31, 40, 42, 43, 45, 46, 57, 64, 65
composite, 33
compression, 16, 31, 34, 47
computation, 21
computing, 67
concentration, 10, 56, 57, 60
configuration, 46, 58, 64
confinement, 1
Congress, iv
conservation, 48
control, 50
controlled, 21
conversion, 3, 42, 43, 44, 45, 46, 52, 55, 57, 60
Coulomb, 5
coupling, 61
cross-phase modulation, 6, 38
crosstalk, 12, 46
crystal, 64
crystalline, 59

D

data transfer, 35
DCF, 64
decay, 12, 55
defects, 6
degenerate, 4, 38, 39, 40, 43, 45, 47
degradation, 3, 60
demand, 34
density, 47, 49, 56, 57
density fluctuations, 47, 56, 57
detection, 3
dielectric, 5, 7
differential equations, 38, 53
diffraction, 1, 47
diodes, 34
dipole, 1, 5, 6
dipole moment, 5, 6
dipole moments, 5, 6
Dirac delta function, 14
dispersion, 2, 3, 4, 12, 13, 14, 15, 16, 17, 18, 23, 24, 25, 26, 28, 29, 31, 32, 33, 34, 35, 38, 39, 40, 41, 42, 45, 46, 47, 48, 51, 62, 63, 64, 65
distribution, 10, 19, 20
division, vii, 3, 33, 34, 62
dopant, 56, 60
doped, 3, 50, 58, 61, 62, 63, 64
doping, 57
Doppler, 48
duration, 27, 36, 49, 60, 63, 65

E

electric field, 5
electromagnetic, 5, 7, 9, 38
electromagnetic fields, 38
electromagnetic wave, 5, 7
electromagnetic waves, 7
electron, 5, 6
electronic, iv, 1, 14, 33, 35, 46, 63
electrostatic, iv
emission, 34, 61, 62
energy, 12, 14, 21, 32, 48, 63, 64
energy transfer, 14
envelope, 15
environmental, 36, 57
erbium, 3, 58, 63, 64
evolution, 15, 16, 26, 27, 29, 30, 31, 52, 55
expert, iv
exponential, 18, 55

F

feedback, 34
feeding, 12
FFT, 18
fiber, vii, 1, 2, 3, 4, 6, 10, 11, 13, 14, 15, 16, 17, 18, 19, 20, 21, 23, 24, 25, 26, 27, 28, 29, 31, 33, 34, 35, 38, 41, 42, 46, 47, 49, 50, 52, 53, 54, 55, 56, 57, 58, 59, 60, 61, 62, 63, 64, 65, 72
fiber Bragg grating, 57, 62
fiber optics, 13
fibers, 1, 3, 4, 6, 10, 11, 27, 28, 38, 46, 48, 49, 56, 60, 62, 63, 64, 65, 67

filters, 63
flexibility, 35, 61
fluctuations, 3, 47, 56
fluorescence, 58
focusing, 1
Fourier, v, 4, 16, 17, 18, 46
FWHM, 15, 28, 32

G

Gaussian, 4, 10, 15, 23, 24, 26, 27, 28, 30, 49
generation, vii, 1, 2, 3, 4, 6, 16, 35, 38, 40, 41, 44, 45, 58, 59, 64, 65
germanium, 10, 61
glasses, 10, 62
growth, 12, 55
guidance, 10, 48, 65

H

handling, 62
hybrid, 58
hyperbolic, 15, 27, 28

I

implementation, 19
India, 69
inelastic, 58
inferences, 45
Information Technology, 69
injury, iv
instabilities, 36
intensity, 1, 2, 6, 8, 9, 10, 11, 12, 13, 23, 25, 53, 55, 59, 63, 65
interaction, 1, 2, 4, 6, 11, 27, 36, 46, 47, 56, 59, 60
interactions, 1, 71
interference, 44, 62
inversion, 6, 8
Islam, 73, 74

L

laser, 1, 33, 34, 35, 50, 58, 59, 62, 63, 64, 65, 71
lasers, 4, 27, 34, 61, 62, 64
lead, 1, 3
lifetime, 49
light emitting diode (LED), 34
linear, 1, 2, 5, 7, 8, 9, 10, 12, 15, 16, 24, 26, 30, 32, 40, 43, 63
literature, 1, 50, 67
losses, 13, 28
low power, 58, 64

M

magnetic, iv, 6
management, 64
mapping, 47
measurement, 46, 57
mechanical, iv, 57
mechanical stress, 57
media, 59
medicine, vii, 65
microwave, 57
mirror, 64
mixing, vii, 1, 2, 3, 4, 6, 20, 35, 38, 40, 43, 44, 45, 46, 47, 62, 65
modeling, 69
modulation, 1, 2, 3, 6, 11, 12, 16, 24, 27, 46, 47, 59, 63, 64
molecular structure, 1
molecules, 47, 58
momentum, 48
motion, 47
multiplexing, 33, 62

N

network, 46
New York, iii, iv
noise, 2, 6, 47, 53, 57, 61, 62, 64
non-destructive, 47

nonlinear, 1, 2, 3, 4, 6, 7, 8, 9, 10, 11, 12, 13, 14, 15, 16, 17, 18, 19, 21, 23, 25, 26, 27, 34, 35, 38, 39, 40, 46, 47, 57, 61, 62, 63, 64, 65, 67, 71
nonlinear optical response, 6
nonlinear optics, 1
nonlinearities, 3, 6, 15, 17, 39
non-linearity, 16, 38
normal, 16, 24, 26, 40, 63, 64
NPR, 63
nuclear, 58
numerical analysis, 4

O

operator, 7, 17, 18, 19
optical, vii, 1, 2, 3, 4, 5, 6, 8, 10, 11, 12, 13, 16, 17, 18, 20, 27, 28, 33, 34, 35, 38, 46, 47, 48, 50, 53, 56, 57, 58, 59, 60, 61, 62, 63, 64, 65, 67
optical fiber, 1, 2, 3, 4, 6, 8, 10, 11, 12, 13, 16, 17, 20, 28, 33, 34, 38, 46, 48, 56, 59, 63, 65, 67
optical properties, 1
optical pulses, 13, 16, 35, 63
optical solitons, 3
optical systems, 57
optical transmission, 63
optics, 1, 5, 63
opto-electronic, 63
oscillation, 58
oscillations, 25, 30, 59
oscillator, 63

P

parabolic, 5
parameter, 4, 10, 11, 14, 15, 24, 28, 65
passive, 16, 35, 57, 64
PCF, 65
performance, 3, 46
periodic, 47, 64
permittivity, 8
perturbations, 15

phase conjugation, 67
phase shifts, 40, 45
phonon, 48, 49, 58
phonons, 49
phosphor, 62
photon, 6, 48, 56, 58
photonic, 3, 4, 46, 57, 62, 64, 65
photonic crystal fiber (PCF), 3, 4, 62, 64
play, 15, 16, 25, 26, 60
polarizability, 58
polarization, 5, 6, 7, 8, 10, 14, 15, 46, 47, 50, 58, 60, 62, 63
polarized, 15
poor, 62
power, vii, 2, 3, 5, 15, 16, 20, 23, 28, 29, 32, 40, 42, 44, 45, 46, 47, 48, 49, 50, 51, 52, 53, 54, 55, 56, 57, 60, 61, 62, 64, 67
powers, 15, 37, 38, 45, 47, 50, 57, 58, 60, 61, 64, 65, 67
preparation, iv
propagation, vii, 1, 4, 5, 7, 10, 11, 12, 13, 15, 16, 17, 18, 20, 23, 24, 25, 26, 27, 28, 29, 31, 34, 35, 36, 38, 40, 43, 53, 55, 56, 60, 61, 63, 65
property, iv
pulse, 2, 4, 11, 12, 13, 14, 15, 16, 17, 18, 21, 23, 24, 25, 26, 27, 28, 29, 30, 31, 33, 34, 35, 36, 47, 49, 60, 63, 64, 65
pulses, 2, 3, 12, 14, 24, 27, 28, 30, 33, 34, 35, 47, 49, 60, 63, 64, 65
pumping, 61
pumps, 61

Q

quadrupole, 6

R

radial distance, 10
radiation, 5
radio, 57
Raman, 1, 2, 3, 4, 6, 14, 15, 58, 59, 60, 61, 62, 75, 76

Index

Raman and Brillouin scattering, 1, 2, 15
Raman fiber amplifiers, 61
Raman scattering, 4, 14, 58, 60, 62
random, 47
range, 6, 10, 11, 50, 62, 65
Rayleigh, 15, 62
red shift, 26
reduction, 28, 29, 34, 46
reflection, 62
refractive index, 1, 2, 6, 7, 8, 9, 10, 13, 25, 28, 47, 49, 56, 57, 58, 64
regeneration, 3
regular, 62
repeatability, 19
research, 64
resolution, 19
resonator, 62

S

safety, 62
saturation, 47
scattered light, 58
scattering, 1, 2, 3, 4, 15, 47, 48, 56, 58, 60
Schrödinger equation, 4, 13, 61
seed, 58
selecting, 46
selectivity, 57
Self, 2, 25, 26, 29
self-phase modulation, 65
semiconductor, 34, 61
sensing, 62
sensors, 57, 62
separation, 35, 40, 42, 43, 44, 45, 61
series, 5
services, iv
shape, 17, 24, 25, 26, 27, 29, 30, 60
shock, 14
sign, 24, 48
signals, vii, 3, 33, 34, 45, 47, 57
signal-to-noise ratio, 60, 62
silica, 4, 6, 8, 10, 28, 49, 59, 62, 64
silica glass, 6, 59, 60
silicate, 62
Singapore, 71

single mode fibers, 10, 38, 56, 62
soliton, 4, 16, 20, 27, 34, 63, 64
solitons, 63, 64
solutions, 53
spatial, 47, 55, 65
spectra, 34
spectroscopy, 63
spectrum, 14, 26, 27, 29, 30, 32, 33, 34, 35, 37, 49, 57, 59, 60, 61
speed, 46
SRS, 6, 59, 60, 61
stability, 34
steady state, 1, 26, 52
strain, 57
strength, 1, 5, 6, 10, 11
stretching, 63
superposition, 49
suppression, 46, 57
susceptibility, 1, 4, 6, 7, 9, 10, 38, 67
switching, 3
symmetry, 6, 56
synchronous, 35
systems, vii, 3, 19, 21, 34, 38, 46, 53, 57, 60, 62, 63, 64, 67

T

Taylor series, 9, 40, 43
technological, 46
technology, 33, 61, 63
telecommunication, 6, 10
telecommunications, 6
temperature, 57
temporal, 2, 25, 58, 60, 64, 65
theoretical, 47
theory, 4, 30, 45, 47, 56
third order, 1, 2, 4, 5, 6, 7, 8, 9, 10, 14, 18, 23, 35, 38, 40, 43, 67
threshold, 3, 4, 50, 51, 52, 55, 56, 57, 60, 62, 64
Ti, 65
time, 6, 11, 13, 19, 21, 23, 25, 26, 27, 28, 29, 30, 31, 33, 35, 36, 49, 58
trans, 3
transfer, 12, 35, 47, 62

transformation, 59
transmission, 2, 3, 20, 33, 34, 35, 38, 46, 61, 62, 64
transparent, 15
travel, 12, 26

U

uniform, 58

V

values, 10, 11, 15, 28, 42, 52, 54, 55, 63
variable, 19, 50
variation, 13, 30, 44, 45, 47, 51, 55, 58
vector, 5
velocity, 2, 12, 13, 20, 47, 49
vibration, 59
vibrational, 14, 23, 58, 59
vibrational modes, 58, 59
video, 57
visible, 65

W

wave packet, 15
wave vector, 48
waveguide, 15, 34
wavelengths, 14, 20, 27, 34, 35, 37, 38, 40, 44, 45, 46, 47, 58, 59, 60, 61, 62, 65
writing, 6, 7

Y

yield, 33, 55
ytterbium, 61